Reviews of
Environmental Contamination
and Toxicology

VOLUME 210

Reviews of Environmental Contamination and Toxicology

Editor

David M. Whitacre

VOLUME 210

Springer

Coordinating Board of Editors

ISSN 0179-5953
ISBN 978-1-4614-2757-5   ISBN 978-1-4419-7615-4 (eBook)
DOI 10.1007/978-1-4419-7615-4
Springer New York Dordrecht Heidelberg London

Foreword

International concern in scientific, industrial, and governmental communities over traces of xenobiotics in foods and in both abiotic and biotic environments has justified the present triumvirate of specialized publications in this field: comprehensive reviews, rapidly published research papers and progress reports, and archival documentations. These three international publications are integrated and scheduled to provide the coherency essential for nonduplicative and current progress in a field as dynamic and complex as environmental contamination and toxicology. This series is reserved exclusively for the diversified literature on "toxic" chemicals in our food, our feeds, our homes, recreational and working surroundings, our domestic animals, our wildlife, and ourselves. Tremendous efforts worldwide have been mobilized to evaluate the nature, presence, magnitude, fate, and toxicology of the chemicals loosed upon the Earth. Among the sequelae of this broad new emphasis is an undeniable need for an articulated set of authoritative publications, where one can find the latest important world literature produced by these emerging areas of science together with documentation of pertinent ancillary legislation.

Research directors and legislative or administrative advisers do not have the time to scan the escalating number of technical publications that may contain articles important to current responsibility. Rather, these individuals need the background provided by detailed reviews and the assurance that the latest information is made available to them, all with minimal literature searching. Similarly, the scientist assigned or attracted to a new problem is required to glean all literature pertinent to the task, to publish new developments or important new experimental details quickly, to inform others of findings that might alter their own efforts, and eventually to publish all his/her supporting data and conclusions for archival purposes.

In the fields of environmental contamination and toxicology, the sum of these concerns and responsibilities is decisively addressed by the uniform, encompassing, and timely publication format of the Springer triumvirate:

Reviews of Environmental Contamination and Toxicology [Vol. 1 through 97 (1962–1986) as Residue Reviews] for detailed review articles concerned with any aspects of chemical contaminants, including pesticides, in the total environment with toxicological considerations and consequences.

Bulletin of Environmental Contamination and Toxicology (Vol. 1 in 1966) for
rapid publication of short reports of significant advances and discoveries
in the fields of air, soil, water, and food contamination and pollution as
well as methodology and other disciplines concerned with the introduction,
presence, and effects of toxicants in the total environment.

Archives of Environmental Contamination and Toxicology (Vol. 1 in 1973)
for important complete articles emphasizing and describing original experimental or theoretical research work pertaining to the scientific aspects of
chemical contaminants in the environment.

Manuscripts for Reviews and the Archives are in identical formats and are peer
reviewed by scientists in the field for adequacy and value; manuscripts for the
Bulletin are also reviewed, but are published by photo-offset from camera-ready
copy to provide the latest results with minimum delay. The individual editors of
these three publications comprise the joint Coordinating Board of Editors with refer-
ral within the board of manuscripts submitted to one publication but deemed by
major emphasis or length more suitable for one of the others.

Coordinating Board of Editors

Preface

The role of *Reviews* is to publish detailed scientific review articles on all aspects of environmental contamination and associated toxicological consequences. Such articles facilitate the often complex task of accessing and interpreting cogent scientific data within the confines of one or more closely related research fields.

In the nearly 50 years since *Reviews of Environmental Contamination and Toxicology* (formerly *Residue Reviews*) was first published, the number, scope, and complexity of environmental pollution incidents have grown unabated. During this entire period, the emphasis has been on publishing articles that address the presence and toxicity of environmental contaminants. New research is published each year on a myriad of environmental pollution issues facing people worldwide. This fact, and the routine discovery and reporting of new environmental contamination cases, creates an increasingly important function for *Reviews*.

The staggering volume of scientific literature demands remedy by which data can be synthesized and made available to readers in an abridged form. *Reviews* addresses this need and provides detailed reviews worldwide to key scientists and science or policy administrators, whether employed by government, universities, or the private sector.

There is a panoply of environmental issues and concerns on which many scientists have focused their research in past years. The scope of this list is quite broad, encompassing environmental events globally that affect marine and terrestrial ecosystems; biotic and abiotic environments; impacts on plants, humans, and wildlife; and pollutants, both chemical and radioactive; as well as the ravages of environmental disease in virtually all environmental media (soil, water, air). New or enhanced safety and environmental concerns have emerged in the last decade to be added to incidents covered by the media, studied by scientists, and addressed by governmental and private institutions. Among these are events so striking that they are creating a paradigm shift. Two in particular are at the center of everincreasing media as well as scientific attention: bioterrorism and global warming. Unfortunately, these very worrisome issues are now superimposed on the already extensive list of ongoing environmental challenges.

The ultimate role of publishing scientific research is to enhance understanding of the environment in ways that allow the public to be better informed. The term "informed public" as used by Thomas Jefferson in the age of enlightenment conveyed the thought of soundness and good judgment. In the modern sense, being "well informed" has the narrower meaning of having access to sufficient information. Because the public still gets most of its information on science and technology from TV news and reports, the role for scientists as interpreters and brokers of scientific information to the public will grow rather than diminish. Environmentalism is the newest global political force, resulting in the emergence of multinational consortia to control pollution and the evolution of the environmental ethic. Will the new politics of the twenty-first century involve a consortium of technologists and environmentalists, or a progressive confrontation? These matters are of genuine concern to governmental agencies and legislative bodies around the world.

For those who make the decisions about how our planet is managed, there is an ongoing need for continual surveillance and intelligent controls to avoid endangering the environment, public health, and wildlife. Ensuring safety-in-use of the many chemicals involved in our highly industrialized culture is a dynamic challenge, for the old, established materials are continually being displaced by newly developed molecules more acceptable to federal and state regulatory agencies, public health officials, and environmentalists.

Reviews publishes synoptic articles designed to treat the presence, fate, and, if possible, the safety of xenobiotics in any segment of the environment. These reviews can be either general or specific, but properly lie in the domains of analytical chemistry and its methodology, biochemistry, human and animal medicine, legislation, pharmacology, physiology, toxicology, and regulation. Certain affairs in food technology concerned specifically with pesticide and other food-additive problems may also be appropriate.

Because manuscripts are published in the order in which they are received in final form, it may seem that some important aspects have been neglected at times. However, these apparent omissions are recognized, and pertinent manuscripts are likely in preparation or planned. The field is so very large and the interests in it are so varied that the editor and the editorial board earnestly solicit authors and suggestions of underrepresented topics to make this international book series yet more useful and worthwhile.

Justification for the preparation of any review for this book series is that it deals with some aspect of the many real problems arising from the presence of foreign chemicals in our surroundings. Thus, manuscripts may encompass case studies from any country. Food additives, including pesticides, or their metabolites that may persist into human food and animal feeds are within this scope. Additionally, chemical contamination in any manner of air, water, soil, or plant or animal life is within these objectives and their purview.

Manuscripts are often contributed by invitation. However, nominations for new topics or topics in areas that are rapidly advancing are welcome. Preliminary communication with the editor is recommended before volunteered review manuscripts are submitted.

Summerfield, North Carolina   David M. Whitacre

Contents

Attenuation of Chromium Toxicity by Bioremediation Technology . . .   1
Monalisa Mohanty and Hemanta Kumar Patra

The Effects of Radionuclides on Animal Behavior   35
Beatrice Gagnaire, Christelle Adam-Guillermin, Alexandre Bouron,
and Philippe Lestaevel

**Illicit Drugs: Contaminants in the Environment and Utility in
Forensic Epidemiology**   59
Christian G. Daughton

Cumulative Subject Matter Index Volumes 201–210   111

Contributors

Christelle Adam-Guillermin Laboratoire de Radioécologie et d'Ecotoxicologie, IRSN (Institut de Radioprotection et de Sûreté Nucléaire), Centre de Cadarache, 13115 Saint-Paul-Lez-Durance Cedex, France, christelle.adam-guillermin@irsn.fr

Alexandre Bouron CEA, Institut de Recherche en Technologies et Sciences pour le Vivant, Grenoble, France; CNRS, UMR 5249, Laboratoire de Chimie et Biologie des Métaux, Grenoble, France; Université Joseph Fourier, Grenoble, France, alexandre.bouron@cea.fr

Christian G. Daughton Environmental Chemistry Branch, National Exposure Research Laboratory, U.S. Environmental Protection Agency, Las Vegas, NV 89119, USA, daughton.christian@epa.gov

Beatrice Gagnaire Laboratoire de Radioécologie et d'Ecotoxicologie, IRSN (Institut de Radioprotection et de Sûreté Nucléaire), Centre de Cadarache, Bat 186, 13115 Saint-Paul-Lez-Durance Cedex, France, beatrice.gagnaire@irsn.fr

Philippe Lestaevel Laboratoire de Radiotoxicologie Expérimentale, IRSN, BP17, 92262 Fontenay-aux-Roses Cedex, France, philippe.lestaevel@irsn.fr

Monalisa Mohanty Laboratory of Environmental Physiology, Post Graduate Department of Botany, Utkal University, Bhubaneswar 751004, Orissa, India, 18.monalisa@gmail.com

Hemanta Kumar Patra Laboratory of Environmental Physiology, Post Graduate Department of Botany, Utkal University, Bhubaneswar 751004, Orissa, India, drhkpatra@yahoo.com

Attenuation of Chromium Toxicity by Bioremediation Technology

Monalisa Mohanty and Hemanta Kumar Patra

Contents

1 Introduction 1
2 Chemistry of Chromium 2
3 Sources of Chromium in the Environment 4
   3.1 Production, Sources, and Uses of Chromium 4
   3.2 Chromium in Fertilizers, Animal Wastes, and Sewage Sludge 5
4 Chromium Transport and Accumulation in Plants 5
5 Chromium Toxicity 7
   5.1 Effects of Chromium on Microorganisms 7
   5.2 Effects of Chromium on Human Health 8
   5.3 Chromium Phytotoxicity 9
6 Technology for Chromium Bioremediation 13
   6.1 Microbial Remediation 14
   6.2 Green Remediation 16
7 Summary 23
References 24

1 Introduction

Human activities, such as industrial and energy production, mineral excavation, and transportation, result in contamination by polluting substances, many of which are dangerous. Chromium (Cr) is one of the most toxic heavy metals and is discharged into the environment through various human activities. Extensive use of chromium in electroplating, tanning, and textile dyeing and as a biocide in power plant cooling

M. Mohanty (✉)
Laboratory of Environmental Physiology, Post Graduate Department of Botany, Utkal University, Bhubaneswar 751004, Orissa, India
e-mail: 18.monalisa@gmail.com

D.M. Whitacre (ed.), *Reviews of Environmental Contamination and Toxicology*,
Reviews of Environmental Contamination and Toxicology 210,
DOI 10.1007/978-1-4419-7615-4_1, © Springer Science+Business Media, LLC 2011

water results in the discharge of chromium-containing effluents. The pace of release of organic pollutants, and Cr in particular, into the environment is growing exponentially and is enhancing concerns that such releases pose potentially serious risks to human health. Heavy metals, such as chromium, are not destroyed by degradation and are therefore accumulating in the environment.

Chromium has received special attention because it is known to be toxic to humans and animals (WHO 1988; ATSDR 2001) and to plants as well (Panda and Patra 1997; Zayed and Terry 2003; Panda and Choudhury 2005; Shanker et al. 2005a; Nayak et al. 2008; Mohanty and Patra 2009). The World Health Organization (WHO 1988) has addressed the toxic threats of chromium (Cr^{+6}) and has listed it as a human carcinogen (ATSDR 2001).

The Cr that contaminates soils, groundwater, and surface waters must physically be removed from many contaminated sites if these sites are to be rendered usable. The traditional solution for dealing with Cr-polluted sites is to shift contaminated soil into landfills. Such a method may be effective, but is expensive and involves exposure risks of its own. An alternative for rendering Cr-contaminated sites suitable for sustainable development may involve use of living organisms. This would typically include plants or microbes that are capable of degrading, absorbing, or otherwise removing toxic materials from the environment; treatment with such organisms would be designed to stabilize tailings (mined ore) and Cr contaminants in situ. Such approaches are called bioremediation, and are not only feasible, but less costly and more environmental friendly than the traditionally used approaches. In most cases, bioremediation relies for its effectiveness on natural processes within the selected organism. The purpose of the present review is to provide insights into how the risks posed by chromium-contaminated sites may be attenuated at different places worldwide, by applying the various tools and techniques of bioremediation.

2 Chemistry of Chromium

Chromium was discovered by N.L. Vauquelin in 1798. This substance is a steel-gray, lustrous, hard, and brittle metal that belongs to Group VIB, the transition series elements. It is the seventh most abundant element on earth and is the 21st most abundant element in crustal rocks (Katz and Salem 1994). Cr exhibits a range of oxidation states (Table 1). Of the many oxidation states possible for chromium, there are two stable forms, i.e., Cr(III) and Cr(VI). The most toxic form of Cr is Cr(VI). The primary physico-chemical properties of chromium are summarized in Table 2. Within the normal ranges of Eh and pH in soil, chromium exists in four states, viz., two trivalent forms (Cr^{+3} and CrO^{2-}) and two hexavalent forms ($Cr_2O_7^{-2}$ and CrO_4^{-2}) (Bartlett and Kimble 1976). At a pH above 4, the solubility of chromium(III) decreases and apparent complete precipitation occurs at pH 5.5 (Bartlett and Kimble 1976). In aqueous systems, chromium exists primarily in two oxidation states, viz., hexavalent chromium (Cr-VI) and trivalent chromium (Cr-III) (Table 3A). Of these, Cr(III) is generally considered to be the more stable

D.M. Whitacre (ed.), *Reviews of Environmental Contamination and Toxicology*,
Reviews of Environmental Contamination and Toxicology 210,
DOI 10.1007/978-1-4419-7615-4_1, © Springer Science+Business Media, LLC 2011

Table 1 Oxidation states of various chemical species of chromium

Chemical species of chromium	Oxidation states	Examples/occurrence	Remarks
Elemental	0	–	Does not occur naturally
Divalent	Cr(II)	$CrBr_2$, $CrCl_2$, $CrFe_2$, $CrSe$, Cr_2Si	Unstable and oxidized to Cr(III) stage
Trivalent	Cr(III)	CrB, CrB_2, $CrBr_3$, $CrCl_3.6H_2O$	Occur in nature as ores such as ferrochromite
Tetravalent	Cr(IV)	$CrCl_3$, CrF_3, CrN, $KCr(SO_4)_2.12\ H_2O$	Does not occur naturally
Pentavalent	Cr(V)	CrO_2, CrF_4	Does not occur naturally
Hexavalent	Cr(VI)	$(NH_4)_2CrO_4$, $BaCrO_4$, $CaCrO_4$, K_2CrO_4, $_2Cr_2O_7$	Rarely occur in nature, most toxic form, produced by human activities

Sources: Nieboer and Jusys (1988) and Katz and Salem (1994)

Table 2 Physico-chemical properties of chromium

Properties	Analytical data
Phase at room temperature	Solid
Color	Silvery white
Atomic number	24
Atomic mass	51.996 g/mol
Electronic configuration	$[Ar]4s^1 3d^5$
Electronegativity	1.66
Density	7.19 g/cm^3 at 20°C
Hardness	9 Mohs
Melting point	1,875°C or 2,130.2 K
Boiling point	2,672°C or 2,963 K
van der Waals radius	0.127 nm
Ionic radius	0.061 nm (+3); 0.044 nm (+6)
Isotopes	6
Energy of first ionization	651.1 kJ/mol
Heat of fusion	15.3 kJ/mol
Heat of vaporization	347 kJ/mol
Heat of atomization	397 kJ/mol
Thermal conductivity	93.9 J/m s K
Electrical conductivity (1mohm/cm)	77.519
Electron affinity	64.3 kJ/mol
Atomic radius	128 pm
Common oxidation numbers	+3, +6, −2, −1
Other oxidation numbers	+1, +2, +4, +5
Standard potential	−0.71 V (Cr^{3+}/Cr)

form (Banu and Ramaswamy 1997). Cr(VI) is not only more toxic than Cr(III), the latter form of the element is also an essential trace element connected with the glucose tolerance factor (Mertz 1969; Saner 1980). The concentration of chromium found in various environmental media (soil, water, air, and living organisms), and its recommended limits in such media and organisms are presented in Table 3B.

Table 3 Concentration of chromium found in the environment and recommended limits in environmental media and in organisms

A Concentration limits in organisms and environmental media

| Sample types | Concentrations (μg/L) | | References |
	Cr(VI)	Cr(III)	
Fresh water life	1	8	Krishnamurthy and Wilkens (1994), Pawlisz (1997)
Marine life	1	50	Krishnamurthy and Wilkens (1994), Pawlisz (1997)
Irrigation water	8	5	Krishnamurthy and Wilkens (1994), Pawlisz (1997)
Drinking water	50	50	Krishnamurthy and Wilkens (1994), Pawlisz (1997)

B Concentrations of chromium found in various environmental media

Sample types	Total 'Cr' concentrations	References
Natural soil	5–1,000 mg/kg	Adriano (1986)
	30–300 mg/kg	Katz and Salem (1994)
Fresh water	0–117 μg/L	Pawlisz (1997)
	Avg-9.7 μg/L	Pawlisz (1997)
Seawater	0–0.5 μg/L	Pawlisz (1997)
Air	1–5,45,000 ngm^3	Pawlisz (1997)
	100 ngm^3	U.S. EPA (1983)
Plants	0.006–18 mg/kg	Pawlisz (1997)
Animals	0.03–1.6 mg/kg	Pawlisz (1997)

3 Sources of Chromium in the Environment

3.1 Production, Sources, and Uses of Chromium

The level of world production of chromium is in the order of 10^7 t/year. In 1998, the production level stood at 3.4 million t. The countries that constitute the major sources of chromite ore, from which Cr is taken, and their proportionate share (%) are as follows: South Africa (36%), USSR (28%), Turkey (7%), India (6.5%), Albania (6%), Finland (5%), Zimbabwe (5%), and trace amounts in other countries. In India, approximately 98% of chromium deposits are located in the state of Orissa, of which 94% fall in the Sukinda mining belt of Jajpur district and the rest (4%) in the Dhenkanal, Balasore, and Keonjhar districts.

Cr is used in stainless steel alloys, which consume between 50 and 70% of total Cr demand. Cr is also used in the chemical industry for leather tanning, pigment production, and electroplating (Stern 1982). Chandra et al. (1997) estimated that, in India alone, 2,000–3,200 t of elemental Cr escape into the environment annually from tanning industry emissions. The ferrochrome industries emit 12,360 t Cr/year. The combustion of fossil fuels, such as coal (Kessler et al. 1971) and petroleum, also results in the release of chromium into the atmosphere. Coal combustion releases 520 t Cr/year. Moreover, chromium is widely distributed in rocks, fresh water, and seawater, and these may serve as natural sources of Cr loss to the environment. Limestone contains traces of Cr of up to 300 mg Cr/kg limestone (McGrath and Smith 1990).

Chromium metal is used mainly for making steel and other alloys (ATSDR 1998). Cr provides additional strength, hardness, and toughness to steel. It also gives corrosion resistance to steel. Stainless steel, high-speed steel, and corrosion and heat-resistant steel are important varieties of chromium steel. Low-Cr steels (less Cr and small amounts of Ni) are used in the rails of railroads, automobiles, cutlery, and cooking utensils. Cr steel includes stainless steels (12–18% Cr) and super stainless steels (12–30% Cr and 7–10% Ni). The former are used to make cutlery and cooking utensils, and the latter are used to make parts for aircraft and high-speed trains. Chromium compounds, either in chromium(III) or chromium(VI) forms, are also used for chrome plating, manufacture of dyes and pigments, leather and wood preservation, and the treatment of cooling tower water. Smaller amounts are used in drilling mud, textiles, and toner for copying machines (ATSDR 1998). Chromite is used in refractory industry (commercial entities that use heat-resistant materials to line the walls of high-temperature furnaces and reactors) due to its corrosion- and high temperature-resistance and its chemically neutral character. The chromite ore, after extraction (ore tailings), is used in the form of lumps, bricks, or cement in linings, especially linings used in steel blast furnaces. Chromite is used to make chromates and dichromates of Na, K, and Cr and pigments such as chromic oxide green and chromic acid. In turn, these pigments are used in Cr-plating solutions.

3.2 *Chromium in Fertilizers, Animal Wastes, and Sewage Sludge*

Fertilizers and animal wastes contain chromium (McGrath and Smith 1990). The chromium content of sewage sludge ranges from 40 to 8,000 ppm (Berrow and Webber 1972). Fly ash from thermal plants that consume coal is often disposed of by distributing it on land, and this constitutes another major source of Cr input to soils. Other sources, which contribute Cr in trace amounts to the environment, are asbestos, brake linings in vehicles, and aerosols produced from Cr catalysts used in emission-reduction systems for treating exhaust fumes.

4 Chromium Transport and Accumulation in Plants

Chromium is similar to other heavy metals (e.g., As, Cd, Co, Cu, Ni, Sn, and Zn) in that it is phytotoxic at a concentration above a certain threshold level (Nieboer and Richardson 1980). Cr as a trace element is not ranked as an 'essential element' for plants (Huffman and Allaway 1973). However, its essentiality for animal nutrition has received considerable attention from those who study the role of plants as Cr transmitters in the food chain.

Chromium is actively transported across biological membranes in both prokaryotes (Dreyfuss 1964) and eukaryotes (Wiegand et al. 1985; Alexander and Ashet 1995). Once taken inside the cell, Cr(VI) is reduced to Cr(III), possibly because of the unstable nature of chromium in intermediate states like Cr(V) and Cr(IV) (Arslan et al. 1987; Liu et al. 1995). Evidently, Cr^{+3} and CrO_4^{-2} enter vascular tissue with difficulty, but once they gain entry, they are readily transported to the xylem. Hexavalent chromium in the form of CrO_4^{-2} moves more readily than does Cr^{+3}, because the latter may be detained by ion exchange interactions on vessel walls (Skeffington et al. 1976).

CrO_4^{-2} is actively transported across membranes with the help of sulfate-containing protein carriers and, in roots, is immediately converted to Cr^{+3}, possibly by an Fe(III) reductase enzyme (Zayed et al. 1998). In contrast, Cr^{+3} is passively absorbed and retained by cation exchange sites on cell walls (Marschner 1995). McGrath (1982) reported that despite the different properties of Cr^{+3} and CrO_4^{-2}, there were no substantial differences in their rates of absorption and uptake. Skeffington et al. (1976) have suggested that Cr(III) uptake does not require metabolic energy (is a passive transport process, i.e., diffusion), whereas the uptake of Cr(VI) ions occurs by active transport mechanisms. Translocation studies in vegetable crops with Cr indicated that CrO_4^{2-} is converted in roots to Cr^{+3} by all plants tested (Zayed et al. 1998), and translocation of Cr from roots to shoots was extremely limited.

Sulfate (SO_4^{-2}) and other Cr(VI) anions are competitive inhibitors of chromate and inhibit its uptake. In contrast, the presence of Ca^{+3} stimulates the uptake of chromate (Shewry and Peterson 1974). Many researchers (Huffman and Allaway 1973; Lahouti and Peterson 1979; Myttenaere and Mousny 1974; Parr and Taylor 1980; McGrath 1982; Zayed et al. 1998) were of the opinion that Cr(III) and Cr(VI) are poorly translocated to the aerial parts (shoots) of plants and tend to be largely retained at sites in the root. Chromium is absorbed by the roots from nutrient solution as Cr^{+3} or CrO_4^{-2}. It has been found that roots accumulate 10–100 times more Cr than do shoots (Zayed et al. 1998; Srivastava et al. 1999; Skeffington et al. 1976).

Transport and accumulation of chromium depends on the formation of complexes that act to enhance Cr uptake and availability in plants (Athalye et al. 1995; Shanker et al. 2005a, b; Torresdey et al. 2005; Zhuang et al. 2007). Complexes are formed with several organic compounds, i.e., oxalic acids, malate, glycine, EDTA (ethylene diamine tetraacetic acid), DTPA (diethylene triamine pentaacetic acid), EDDHA (ethylene diamine di-ortho hydroxy phenylacetic acid), etc. When complexed with different compounds, uptake rates of Cr^{+3} and Cr^{+6} varied (Athalye et al. 1995; Shanker et al. 2005a; Zhuang et al. 2007). Metabolic inhibitors such as sodium azide and dinitrophenol (DNP) substantially reduced uptake of Cr^{+6}. Alternatively, Cr^{+3} uptake was not affected by metabolic inhibitors in barley seedlings (Skeffington et al. 1976). In several plants (tomato, wheat, potato, bean, pea, beet, barley, maize, spinach, etc.), Cr uptake was enhanced when supplied as Cr(III), Cr(VI), Cr-oxalate, Cr-tartarate, Cr-EDTA, Cr-DTPA, Cr-methionine, or Cr-citrate (Cary et al. 1977a; Athalye et al. 1995; Erenoglu et al. 2007). Salicylic acid complexed with chromium substantially increased the uptake of Cr (Tripathi and Chandra 1991). This may result from the fact that Cr-EDTA, Cr-DTPA, or other complexed chromium compounds are not retained or impeded by ion exchange interactions with the cell walls (Myttenaere and Mousny 1974; Athalye et al. 1995; Cary et al. 1977a).

Translocation and accumulation of chromium depends on the following major factors: the oxidation state of chromium (Mishra et al. 1995) and the concentration of chromium in the growth medium (Kleiman and Cogliatti 1998) or concentration in the plant (Zayed et al. 1998). The experimental results also indicated that Cr accumulation was comparatively higher in plants supplied with CrO_4^{-2} and Cr^{+6} than in plants supplied with Cr^{+3}. The high concentrations of Cr in the nutrient medium also led to an increased accumulation of Cr in plants (Zayed et al. 1998; Mishra et al. 1995). Vegetable crops, *Brassica* spp., sulfur-loving species (cauliflower, cabbage, and kale), spices, and many other crops have the ability to accumulate more Cr in roots than in other plant parts. Water hyacinth is known as a hyperaccumulator of Cr (Lytle et al. 1998; Zhu et al. 1999).

5 Chromium Toxicity

The biological effects of Cr toxicity have been studied and reviewed by many workers (Zayed et al. 1998; Zayed and Terry 2003; Skeffington et al. 1976; Srivastava et al. 1999; Zhu et al. 1999; McGrath 1982; McGrath et al. 1997; Panda and Patra 1997, 2004; Nayak et al. 2004; Erenoglu et al. 2007). Chromium contamination is known to affect organisms in the biosphere at many locations worldwide (Cunningham et al. 1997; Raskin and Ensley 2000; Meagher 2000). Excess concentrations of several heavy metals, including Cr(VI), have resulted in the disruption of both natural aquatic and terrestrial ecosystems (Gardea Torresdey et al. 1998; Meagher 2000). The increase of Cr levels in soil and water has been reported to cause adverse effects to microflora and growing plants. These adverse effects are addressed below.

5.1 Effects of Chromium on Microorganisms

The toxic and mutagenic effects of chromates on microbes are also well documented. The toxic effects of Cr on bacteria and algae have been reviewed by Wong and Trevors (1988). Mertz (1969) reported that chromium is a component of the electron transport chain located inside the plasma membrane of prokaryotes. Cr(VI) acts as a terminal electron acceptor as does oxygen. Ross et al. (1981) found that 10–12 mg/L of Cr(VI) was inhibitory to soil bacteria in liquid media, whereas Cr(III) at this concentration had no effect. Ajmal et al. (1984) reported that chrome-electroplating waste was toxic to saprophytic and nitrifying bacteria, and showed increasing toxicity as the Cr(VI) content of the waste increased. *Rhizobium* has also been observed to be very sensitive to Cr (Misra et al. 1994, 2004; Thatoi 1994; Patnaik 1995; Mishra 2002).

The in vivo generation of Cr(V) from Cr(VI) by *Spirogyra* and *Mougeotia* has been reported by Liu et al. (1995). Growth of *Scenedesmus acutus* was detected at concentrations of Cr exceeding 15 ppm (Travieso et al. 1999). However, Brady et al. (1994) reported that colonial algal growth of *Scenedesmus* and *Selenastrum* was possible at levels of 100 ppm of Cr(III), but not at 100 ppm of Cr(VI). A lengthening in the lag growth phase was induced by Cr(VI), whereas the growth rate was decreased by Cr(III), in *Euglena gracilis* (Brochiero et al. 1984). Cr(VI) also induced an alteration in the cytoskeleton, which may have resulted in the loss of motility (Bassi and Donini 1984). Inhibition of photosynthesis by Cr has also been reported in *Chlorella* (Wong and Trevors 1988) and in *Scenedesmus* (Corradi et al. 1995). In estuarine algae, Cr(VI) toxicity is inversely proportional to salinity (Frey et al. 1983).

In *Saccharomyces cerevisiae*, chromium toxicity was stronger in cells grown in non-fermentable substrates (Henderson 1989). Other effects included inhibition of oxygen uptake (Henderson 1989) and induction of *petite* mutations. These results suggest that chromate specifically targets the mitochondria of *S. cerevisiae* (Henderson 1989). Additional effects of Cr in *S. cerevisiae* include gene conversion (Kharab and Singh 1985; Galli et al. 1985) and induction of mutations (Kharab and Singh 1985; Galli et al. 1985; Cheng et al. 1998).

5.2 Effects of Chromium on Human Health

Chromium pollution is known to induce respiratory and skin diseases and affect mucous membranes (ATSDR 1998; USEPA 1998; WHO 1988). Skin ulcers that do not heal are caused by chromium exposure. Such respiratory and skin diseases were frequently found in workers at chromite mining and processing sites. The Cr exposure also affects individuals in the nearby villages. Gastrointestinal and neurological effects have been noted after inhalation exposure to high concentrations of chromium(VI). Dermal exposure to Cr(VI) also induces skin burns in humans (ATSDR 1998; USEPA 1998; WHO 1988). Chronic inhalation exposure to chromium(VI) in humans results in respiratory tract effects, such as perforations and ulcerations of the septum, bronchitis, decreased pulmonary function, pneumonia, asthma, nasal itching, and soreness (ATSDR 1998; USEPA 1998; WHO 1988). Workers exposed to chromium(VI) compounds may be at risk for developing cancer. Based on results of animal studies, EPA has concluded that only chromium(VI) should be classified as a human carcinogen (ATSDR 1998; USEPA 1999).

There is only limited information available on the reproductive effects of chromium(VI) in humans exposed by the inhalation route; this information suggests that exposure to chromium(VI) may result in complications during pregnancy and childbirth (ATSDR 1998). Chromium(III) is an essential element in humans and has a recommended daily intake of 50–200 µg/day for adults (ATSDR 1998). Acute oral animal tests have shown chromium(III) to have moderate toxicity (ATSDR 1998; USDHHS 1993).

5.3 Chromium Phytotoxicity

Chromium is mainly present in the environment in two stable forms i.e., Cr(III) and Cr(VI). The activities of these two forms markedly differ because of their different abilities to cross biological membranes (Debatto and Luciani 1988). Chromium(VI), as chromate, readily penetrates plant cuticle and membranes via a general anion transport system, whereas Cr(III) complexes do not diffuse through plant membranes. The accumulation of chromium(III) at the cell surface results from cation binding activity.

Several workers have reported chromium toxicity in plants (Koenig 1910; Lyon et al. 1970; Wallace et al. 1976; Watanabe 1984; Moral et al. 1993; Misra et al. 1994, 2004). Such toxic effects of chromium includes stunted growth, chlorosis, reduced crop yield, delayed germination, senescence and premature falling of leaves, biochemical lesions, reduced enzyme activity, and reduced synthesis of proteins, amino acids, and enzymes such as RNAse, invertase, amylase, catalase, peroxidase, and Fe-reductase. The phytotoxic effects of chromium were first reported a century ago by Koening (1911). Other effects from uptake of chromium by plants include reduced rates of growth, damage to cell walls and membranes, and changes to plant metabolic status (Williamson and Johnson 1981; Panda and Patra 1997; Nayak et al. 2004; Mohanty et al. 2005). The visual symptoms of Cr toxicity in plants include stunted growth, poorly developed root system, and curled and discolored leaves (Hunter and Vergnano 1953; Misra et al. 1994, 2004). Corradi and Bianchi (1993) also reported suppression of lateral roots as a symptom of chromium toxicity. However, Gaw and Soong (1942) saw reduction in the dry weight and nodulation of peas after adding chromic sulfate to the soil in which they grew. Chlorotic bands on cereals were noted by several researchers (Kabata-Pendias and Pendias 1992; Panda and Patra 1997, 2004; Patra et al. 2002; Nayak et al. 2004), and the exposures that produced such effects resulted in yield reduction (Parr and Taylor 1982; Misra et al. 1994). Immediate wilting and plant death have also been reported from exposure to very high levels of Cr (Parr and Taylor 1982). It has been reported (Austenfeld 1979; Bassi et al. 1990; Bonet et al. 1991; Choudhury and Panda 2005; Mohanty et al. 2005, 2008, 2009; Nayak et al. 2008) that high concentrations of Cr in plants (rice, wheat, lentil, green gram, pistia, lemna, moss, and beans) resulted in stunted growth, reduced chlorophyll content, higher activity of certain enzymes, and higher bioaccumulation of Cr in roots.

5.3.1 Inhibition of Germination and Seedling Growth

High levels of chromium may inhibit seed germination and subsequent seedling growth. Cr(VI) concentrations above 2 mM can affect pea seed germination and suppress the growth of radicle and plumule (Bishnoi et al. 1993). Chromium toxicity causing inhibition of seed germination and radicle growth in plants was also observed by others (Atta Aly et al. 1999; Corradi and Bianchi 1993; Liu et al. 1993; Nayari et al. 1997; Panda et al. 2002). The germination and growth of bush bean were substantially affected at 500 mg/kg by Cr in soil (Parr 1982). Chromium toxicity that inhibits plant growth results from inhibition of cell division through induction of chromosomal aberrations.

5.3.2 Growth Retardation

Cr(VI) is known to produce serious damage in living plant cells, but Cr(III) is less toxic because of its extremely low solubility, which prevents leaching into ground water or uptake by plants. At 100 μM of Cr(III), 40% growth retardation occurs, whereas Cr(VI) showed 75% inhibition in shoots and 90% in roots of barley seedlings. At low 'Cr' concentrations, the dry matter content and shoot and root length were found to increase (Bonet et al. 1991). It was also noted that marked decreases in shoot/root ratio resulted from increasing concentrations of chromium (Cary et al. 1997b; Zayed et al. 1998; Patra et al. 2005).

Visual and other symptoms of Cr toxicity in plants are stunted growth, poorly developed root system, curled and discolored leaves (Pratt 1966), chlorosis and narrow leaves (Hunter and Vergnano 1953), chlorotic bands on cereals (Kabata-Pendias and Pendias 1992), and yield reduction (Hara et al. 1976; Hara and Sonoda 1979; Parr and Taylor 1982). Chromium has deleterious effects on plant growth also because it causes perturbations in mineral nutrition. Cr also causes wilting and plasmolysis in root cells (Bassi et al. 1990; McGrath 1995). After exposure to high Cr concentrations, some plants may exhibit brownish red leaves that display small necrotic areas, purpling of basal tissue (Adriano 1986; Hunter and Vergnano 1953), immediate wilting, and plant death (Hara and Sonoda 1979; Parr and Taylor 1982).

Shewry and Peterson (1974) observed that the first toxic effect of Cr(VI) to plants was inhibition of root and shoot growth. Hauschild (1993) has provided the sequence of observed symptoms after Cr exposure in plants: induction of stress compounds (putrescine and chitinase) > inhibition of root growth >visible damage symptoms > leaf growth.

5.3.3 Photosynthetic Inhibition

A symptom of chromium toxicity in plants is induced disorganization of ultrastructure of chloroplast membranes (Sarkar and Jana 1987; Poschenrieder et al. 1991). Chromium toxicity in plants leads to diminished photosynthesis (Austenfeld 1979; Dubey and Rai 1987) accompanying reduced chlorophyll synthesis (Vazquez et al. 1987). This phenomenon results in visual symptoms in plants such as chlorosis, necrosis, and stunted growth (Panda and Patra 2000a; Panda et al. 2003). The inhibition of photosynthesis was due to ultrastructural changes in the chloroplast, viz., severe decreases in granal and stromal lamellae and swelling of thyllakoid membranes (Poschenrieder et al. 1991; Bassi et al. 1990; Choudhury and Panda 2004). Such effects are observed in several plants such as *Lemna minor, Pistia* spp., *Taxithelium nepalense*, and bean plants. Chromium also causes a decrease in the Hill reaction of chloroplasts and this affects both dark and light reactions (Krupa and Baszynski 1995; Zeid 2001).

5.3.4 Oxidative Stress

Chromium toxicity results in oxidative stress in plants. Such oxidative stress results from the generation of free radicals or reactive oxygen species (ROS) such as O_2^-, H_2O_2, $^{\bullet}OH$ (Panda and Patra 2000b; Panda and Choudhury 2005). These ROS may produce oxidative damage to biological membranes (caused by lipid peroxidation). The basic cause of chromium toxicity emanates from the process of reduction of Cr(VI) to lower oxidation states (Kawanishi et al. 1986), in which free radicals are generated (Kadiiska et al. 1994; Panda et al. 2003; Choudhury and Panda 2005). The formation of a transient form of Cr(V), during the reduction of Cr(VI) to Cr(III), is thought to be one probable mechanism by which ROS are developed and cause their effects in plants (Kawanishi et al. 1986). The entities responsible for the formation of Cr(V) from Cr(VI) are physiological reducing agents such as NAD(P)H, $FADH_2$, several pentoses, and glutathione (Shi and Dalal 1989). Cr(V) complexes react with H_2O_2 to generate significant amount of $^{\bullet}OH$ radicals, which may directly trigger interactions with DNA and/or induce other toxic effects.

5.3.5 Enzymatic Changes

Several antioxidant enzymes (catalase, peroxidase, glutathione reductase, ascorbate peroxidase, and superoxide dismutase) may scavenge the ROS formed by the presence of Cr. These antioxidants mitigate oxidative damage and reduce stress phenomena. The activity of peroxidase and catalase has been studied in response to chromium toxicity in several plant species: rice, wheat, green gram, and in lower plants like mosses (Panda and Patra 1998; Panda 2003). Results show that the activities of these enzymes may be suppressed or induced. Suppressed enzyme activity was observed in plants grown under conditions of toxic levels of Cr, but simultaneously, the synthesis of other enzymes was stimulated. The suppression and induction vary with the plant species involved. Increased synthesis was observed in catalase and peroxidase activity at 100 μM of Cr(VI), in hydroponically grown wheat seedlings. Cr(III) may also react with the carboxyl and sulfhydral groups of enzymes and cause alternation of their structure and activities (Bianchi and Levis 1977; Mohanty et al. 2005, 2008). Although there are several reports of the hyperactivity of antioxidative enzymes in various plants that are under Cu, Pb, Zn stress (Ali et al. 2003; Van Assche and Clijsters 1990), there are few reports available on the role of enzymatic antioxidant systems in protecting plants from ROS stress induced by Cr. This demonstrates the hypothesis of Dong et al. (2007), which is that although antioxidants may alleviate Cr-induced stress, antioxidants may also be a sensitive target of Cr toxicity in plants.

High concentrations of Cr cause protein degradation and inhibit nitrate reductase activity, in some plants (Solomonson and Barber 1990; Vajpayee et al. 1999, 2000). A decline in the content of amino acids such as cysteine (Vajpayee et al. 2001) and increased synthesis of proline were reported to occur at toxic levels of chromium. Moreover, a severe inhibition of cytochrome oxidase activity resulted from the binding of Cr to complex-IV and Cyt-a_3 in the mitochondrial electron transport system

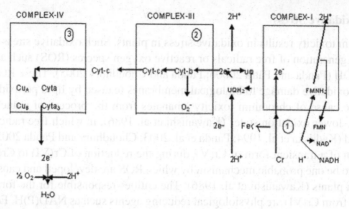

Fig. 1 The electron transport system of plants as influenced by chromium. The *circled numbers* (1, 2, and 3) identify the biochemical sites of action at which chromium may generate its effects (Panda and Choudhury 2005)

(Dixit et al. 2002). The sites in the electron transport chain that may be affected by chromium are depicted in Fig. 1.

Fe-reductase activity was decreased in response to chromium in Fe-deficit plants (Schmidt 1996). Chromium possesses the capacity to degrade δ-amino levulinic acid dehydratase, which is an important enzyme for chlorophyll biosynthesis (Vajpayee et al. 2000).

5.3.6 Macromolecular Damage

Chromium toxicity that produces damage to macromolecules has been observed in several studies. Chromium may inhibit cell division and induce chromosomal aberrations (Liu et al. 1993). Cr(III) and H_2O_2 cause breakage of DNA strands at a pH of 6–8 (Strile et al. 2003). Cr also hampered the replication of DNA by inhibiting transcriptional and translational processes. Binding of chromium ions to DNA molecules is well documented to occur in mammalian (Levis et al. 1975) and hamster cells (Bianchi and Levis 1977). Intracellular Cr(III) may be sequestered by DNA phosphate groups that affect replication and transcription, and may cause mutagenesis (Costa 1997). Cr(III) causes modification of DNA polymerase and other enzyme activities by displacing Mg ions.

5.3.7 Other Phytotoxic Effects

Cr(VI) depresses plant uptake of potassium and impedes H^+ extrusion coupled to K^+ uptake across the plasma membrane, and may produce a decrease in the proton gradient and depolarize transmembrane electric potential energy (Zaccheo et al. 1982; Marre et al. 1974). This energy-linked process is believed to be important in regulating certain important physiological processes such as seed germination and stem elongation (Marre 1979). Chromium causes perturbation in the structure of the

Euglena cell nucleus (Fasulo et al. 1983) and in the trifoliate leaves of bush bean (Vazquez et al. 1987).

5.3.8 Plant Response to Heavy Metals

Plants have three basic strategies for growing in metal-contaminated soil. To protect themselves from the toxic effects of heavy metals, plants may either be (a) excluders, (b) indicators, or (c) hyperaccumulators. (a) *Metal excluders* act to prevent metal from entering their aerial plant parts, or they maintain low and constant metal concentrations as they grow in soils having a broad range of metal concentrations. Such plants store absorbed metals primarily in their roots. These plants may alter their membrane permeability, change metal-binding capacity of cell walls, or exude more chelating substances. (b) *Metal indicators* are species which actively accumulate metal in their aerial tissues at levels that generally reflect metal levels in the soil. They tolerate the existing concentration level of metals by producing intracellular metal-binding compounds (chelators) or alter metal compartmentalization patterns by storing metals in non-sensitive parts. (c) Some plant species can concentrate metal in their aerial parts to levels that far exceed soil levels. These are called *hyperaccumulators*. Such plants can absorb high levels of contaminants and concentrate them either in their roots, shoots, and/or leaves (Fig. 2).

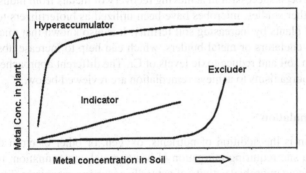

Fig. 2 Conceptual response strategies of metal concentrations in plant tops in relation to increasing total metal concentrations in the soil (Ghosh and Singh 2005a). Notes: *Hyperaccumulators*: The plants that can absorb high levels of contaminants concentrated either in their roots, shoots, and/or leaves. *Indicators*: The tolerant plant species which actively accumulate metal in their aerial tissues and generally reflect metal availability in the soil. *Excluders*: The plants which prevent metal from entering their aerial parts and restrict metal in their roots

6 Technology for Chromium Bioremediation

Heavy metals occur naturally in soils and plants and are integral components of the biosphere. Soil contamination with heavy metals caused by human activities is normally dealt with by transfer of contaminated soil to landfills. However, this approach is expensive and causes its own problems. Therefore, eco-friendly technologies that offer low-cost alternatives are increasingly being sought to replace the former processes. Bioremediation may constitute such an alternative for the future. Bioremediation is the process of using living organisms, typically plants or microbes, to remove toxic elements from the environment (Kumar et al. 1995; Adler 1996; Cunningham and Ow 1996; Negri and Hunchman 1996; Yang et al. 1996; Kiling 1997). Usually, bioremediation takes advantage of natural processes that already exist within selected organisms. The advantages of in situ bioremediation (bioremediation that takes place at the site of the contamination) are that it is cheaper and it eliminates the need to extract or remove the contaminants, thus resulting in reducing the prospect of exposure to workers. The disadvantages are that the site of bioremediation is not contained, and it is harder to control conditions and monitor progress.

Some bioremediation techniques are described below in greater detail. They are broadly classified into two categories depending on the type of organisms used for remediation.

6.1 Microbial Remediation

Many trials and studies have shown that both prokaryote and eukaryote microorganisms have been successful in achieving recovery of metals from industrial waste streams. In other studies, microbes have been utilized as biofertilizers to stimulate the growth of plants by increasing soil fertility. It is also known that microbes may release metal chelators or metal binders, which can help to increase absorption of minerals from soil and reduce toxic levels of Cr. The different approaches that have utilized microorganisms to achieve remediation are reviewed below.

6.1.1 Biostimulation

Biostimulation is the addition of nutrients, oxygen, or other electron donors and acceptors to a site requiring mitigation of heavy metal contamination, for the purpose of enhancing microbial activity of naturally occurring organisms (Leung 2004). Biostimulation provides the basis for the requirement of retaining viable native populations of specific contaminant-degrading microbes that already exist at the site. Successful biostimulation may require amendment or support to achieve the correct environment for accomplishing the degradation or detoxification of a contaminant to an acceptable regulatory level, within a reasonable time period (Quagraine et al. 2005).

6.1.2 Bioaugmentation

Bioaugmentation is the addition of microorganisms, to those that already exist at a site, that can biotransform or biodegrade contaminants (Quagraine et al. 2005). The added microorganisms may be a completely new species or simply more members of a species that already exist at the site. A necessary prerequisite for an efficient bioaugmentation would be the presence of nutrients that can stimulate the growth and activity of these "foreign" microorganisms (Quagraine et al. 2005). But the application of this technique for removal of Cr is not currently in regular practice. This method does not currently offer any obvious treatment advantages, in consumption of time and money. Moreover, stimulating indigenous microbial populations does not constitute, at this time, an acceptable treatment of contaminated sites. One concept for the future is to inoculate local microbial cultures with selected foreign microbes that have demonstrated a capability to degrade the contaminants present. Once acclimatized to their new environment, such innoculants may be entertained for use in bioaugmentation. However, there is the need to examine whether these "foreign" microorganisms that are added to tailings in pond water can compete with the indigenous populations or not (Quagraine et al. 2005).

6.1.3 Bacteria

Bacteria are generally the most commonly used organism for bioremediation. However, fungi, algae, and plants have also been used. Bioengineering of bacterial heavy metal resistance genes has produced biosensors for several toxic metals (Ramanathan et al. 1997). Some bacteria have the ability to reduce Cr(VI) and may be ideal for bioremediation of chromate-polluted areas (Ohtake and Silver 1994; Wakatasuki 1995; Silver and Williams 1984; Loveley et al. 1993). Precipitation of Cr has been reported in anaerobic *Clostridium* and by sulfate-reducing bacteria (Dvorak et al. 1992).

Adsorption of Cr(III) can occur by the heat-dried biomass of the cyanobacterium *Phormidium laminosum* (Sampedro et al. 1995). *Cladophora* accumulated several heavy metals, and this genus accumulated more Cr at a faster uptake rate (72% after 15 min) than it did other metals (Vymazal 1990). Cr(III) was effectively removed (83–99%) in laboratory tests with *Scenedesmus, Selenastrum,* and *Chlorella,* whereas Cr(VI) was rather poorly removed (18–22%) (Brady et al. 1994). The removal of Cu, Ni, Al, and Cr from acidic mine wastes by the red algae *Cyanidium caldarium* occurs by cell surface precipitation of metal-sulfide microcrystals (Wood and Wang 1983).

6.1.4 Yeast and Filamentous Fungi

These organisms offer a viable alternative for the bioremediation of soils polluted by Cr (Cervantes et al. 2001). A new siderophore, rhizoferrin, has been identified in *Mucorales* that shows increased Cr(III) biosorption (Pillichshammer et al. 1995). Chemically treated mycelia from *Mucor mucedo* and *Rhizomucor miehei* efficiently bind Cr (Wales and Sagar 1990). The biomasses obtained from *Rhizomucor arrhizu, Candida tropicalis,* and *Penicillium chrysogenum* were excellent biosorbents of Cr (Volesky and Holan 1995). *S. cerevisiae* and *Candida utilis* have the ability to sorb Cr(VI) and the sorption capacity of dehydrated cells is considerably higher than that of intact cells (Rapoport and Muter 1995). In most cases, Cr accumulation in the chromate-resistant fungi was lower than in chromate-sensitive strains, but the biosorption and bioaccumulation processes were similar (Czako-Ver et al. 1999). The biosorption ability of chromate-resistant mutants could be combined with their ability to reduce chromate. Chromate-resistant strains of *Aspergillus* spp. (Paknikar and Bhide 1993) and *Candida* spp. (Ramirez et al. 2000), isolated from Cr-polluted environments, have shown Cr(VI) reducing activity. Moreover, there is evidence that the heavy metal-tolerant arbuscular mycorrhizal fungi (AM fungi) could protect plants against the harmful effects of excessive heavy metals.

In heavy metal-contaminated sites, AM fungi improve plant growth and survival by increasing plant access to relatively immobile minerals such as P (Vivas et al. 2003; Misra et al. 2004), and this improves soil texture by binding soil particles into stable aggregates that resist wind and water erosion (Rilling and Steinberg 2002). Moreover, AM fungi are capable of binding heavy metals into roots in ways that restrict their translocation into shoot tissues (Dehn and Schuepp 1989; Kaldorf et al. 1999; Misra et al. 2004). AM fungi play a vital role in metal tolerance (del Val et al. 1999) and accumulation (Jamal et al. 2002), because they store a greater volume of metals in their mycorrhizal structures and in their roots and spores. Several heavy metal-tolerant AM fungi have been isolated from polluted soils. These fungi can be useful for reclamation of such degraded soils, because they are naturally associated with many plant species in heavy metal-polluted soils (Gaur and Adholeya 2004).

Increased root/shoot Cr ratio in AM plants has also been found by Misra et al. (2004) in chromite-mine overburden soil. The change in this ratio may point to other mitigation mechanisms, such as dilution by increased root or shoot growth, exclusion by precipitation into poly-phosphate granules, and compartmentalization (Kaldorf et al. 1999; Turnau et al. 1993). Indirect mitigation mechanisms may occur and include the effect of AM fungi on rhizosphere characteristics such as changes in pH (Li et al. 1991), microbial communities (Olsson et al. 1998), and root-exudation patterns (Laheurte et al. 1990). The use of dead fungal biomass of *Aspergillus niger* for the detoxification of Cr(VI) from contaminated waters has also been studied. Park et al. concluded that the mechanism of Cr(VI) removal by dead fungal biomasses such as *A. niger* was a redox reaction. Cr(VI) was reduced to Cr(III) through both direct and indirect mechanisms. The Cr(VI) removal rate was increased with decreased solution pH and with increased Cr(VI) concentration, biomass concentration, and temperature. Dead fungal biomass is abundant, cheap, does not require a continuous nutrient supply for maintaining the cells in good physiological conditions, and dead cells are not subjected to physiological constraints such as metal toxicity.

6.2 Green Remediation

Phytoremediation or "green" remediation is defined as the use of green plants to remove pollutants from the environment or to render them harmless. The generic term 'phytoremediation' consists of the Greek prefix phyto (plant), attached to the Latin root 'remedium' (to correct or remove an evil). This technique is a cost-effective plant-based approach for removal of heavy metals from soil and groundwater (Jena et al. 2004). The success of phytoremediation or phyto-mining depends on the availability of plant species – ideally those native to the region of interest – that are able to tolerate and accumulate high concentrations of heavy metals (Baker and Whiting 2002).

There is currently considerable interest in the use of phytoremediation technology to deal with the problem of chromium- and other heavy metal-contaminated soils, sediments, and water. Although metal-contaminated soil can be remediated by chemical, physical, and biological techniques, the most appropriate technique may be determined by studying the particular category of contamination. Remediation of metal-contaminated soil can be grouped into two categories as defined below (Blaylock and Huang 2000; Cooper et al. 1999; Ghosh and Singh 2005a; Huang et al. 1997).

6.2.1 *Ex Situ* Methods

Ex situ methods require removal of contaminated soil for treatment either on or off site, and then returning the treated soil to the restored site. The conventional *ex situ* methods that are applied to remediate polluted soils relies on excavation and detoxification (physical or chemical destruction). Such treatments may either destroy or may result in the contaminant being solidified or otherwise immobilized. In addition, incineration of contaminants is sometimes used to effect virtual total destruction. The conventional ex situ technique is to excavate heavy metal-contaminated soil and then rebury it at a landfill site (McNeil and Waring 1992; Smith 1993). Such offsite burial of contaminated media is often inappropriate because it merely shifts the contamination problem to a new site; moreover, there are hazards associated with the transport and redeposition of contaminated soil.

6.2.2 *In Situ* method

In the in situ method, remediation occurs without excavation of a contaminated site. Reed et al. (1992) defined in situ remediation technologies as destruction or transformation of the contaminant, immobilization to reduce bioavailability, and separation of the contaminant from the bulk soil. The use of microbial biomass and bioaccumulators helps to capture metals by extracellular precipitation and subsequent intracellular accumulation; thus, the toxic metal ions are immobilized at the site of contamination which reduces their bioavailability. In situ techniques are favored over ex situ techniques because they are cheaper and are still effective in reducing ecosystem impact. Diluting the heavy metal content of a substrate to a safe level by importing it and mixing it with clean soil is sometimes an alternative for on-site management (Musgrove 1991). On-site containment, with appropriate barriers, provides another alternative that involves covering the soil with inert material.

Although the concept has been informally employed for at least 300 years, the modern concept of using metal-accumulating plants to remove contaminating compounds was first introduced in 1983. This technology can be applied to both organic and inorganic pollutants that are present in soil (solid substrate), water (liquid substrate), and air. The technique involves the ability of plants to absorb and concentrate elements of heavy metals from contaminated environmental media (soil and water); metals that are candidates for successful uptake by plants includes Se, Hg, Pb, Cr, Cd, Zn, and Fe. There are five categories of phytoremediation techniques: phytoextraction using hyperaccumulator plants, phytovolatilization, rhizofiltration, phytostabilization, and phytodetoxification (Salt et al. 1995, 1998). These processes will be discussed in detail below.

Phytoextraction

Phytoextraction involves using plants that are capable of accumulating metals from contaminated soils, sediments, and water, at high concentrations, into their tissues (Peterson 1975). It is the best approach to remove and isolate soil contaminates without destroying soil structure and fertility. It is also referred to in the literature as phytoaccumulation (Fig. 3) (USPAR 2000). The selected plant absorbs, concentrates, and precipitates the toxic metals and radionuclides from contaminated soils (Brooks et al. 1998) in their biomass. This technique is best suited for the remediation of areas that are diffusely polluted at relatively low concentrations and where contamination rests primarily near the surface (Rulkens et al. 1998). Several approaches have been used, but the main two strategies for phytoextraction are

(i) Chelate-assisted phytoextraction or induced phytoextraction, in which artificial chelates are added to soil to increase the mobility and uptake of metal contaminant.
(ii) Continuous phytoextraction, in which the removal of metal depends on the natural ability of the plant to remediate; only the number of plant growth repetitions is controlled.

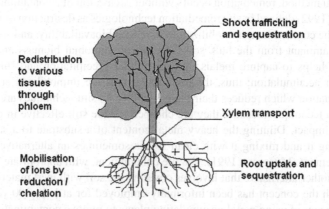

Fig. 3 Pictorial representation of the process of phytoextraction

Hyperaccumulator plants have been used to boost the effectiveness of this technology. To make this technology feasible, the plants must extract large concentrations of heavy metals into their roots, translocate the heavy metals to surface biomass, and produce a large quantity of plant biomass. The removed heavy metals can be recycled through phyto-mining (Nicks and Chamber 1994; Ghosh and Singh 2005a) to produce bio-ore, which is a form of concentrated metal that is produced from the contaminated plant biomass and may be sold (Fig. 4). Factors such as plant growth rate, element selectivity, resistance to disease, and method of harvesting are also important factors in selecting candidate plants as hyperaccumulators. Factors like slow growth, shallow root systems, small biomass production, or difficulty in final disposal limit the use of hyperaccumulator species. Plants such as *Ipomoea carnea, Datura innoxia, Phragmites karka, Cassia tora, Lantana camara, Brassica juncea, Brassica campestris, Leersia hexandra, Convolvulus arvensis, Albizia amara, Cynodon dactylon,* and *Pluchea indica* are being studied for Cr hyperaccumulation capacity by several workers (Torresdey et al. 2004; Shanker et al. 2005b; Ghosh and Singh 2005b; Sampanpanish et al. 2006; Zhuang et al. 2007). Some plants can accumulate remarkable levels (100- to 1,000-fold the levels normally found in most species) of heavy metals. This striking phenomenon, known

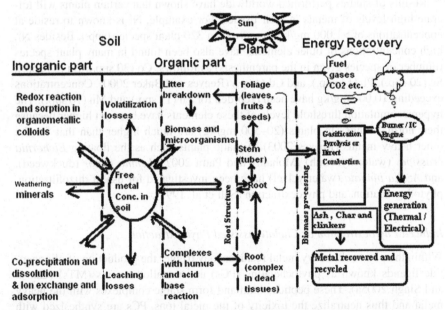

Fig. 4 The soil, plant, and energy recovery system depicting the key components concerned with the mass transfer and dynamics of phytoextraction (Ghosh and Singh 2005a). The soil system involves the generation of free-metal concentration in soil from its inorganic and organic fractions. The plant system extracts (phytoextraction) large concentrations of heavy metals into their roots, stems, and foliage and concentrates these metals in the biomass fraction. The energy recovery system recovers heavy metals from plant biomass for production of bio-ore and generates energy for human needs by thermo-chemical conversion of plant biomass

as metal hyperaccumulation (i.e., the ability to accumulate at least 0.1% of the leaf dry weight as a heavy metal) (Rajakaruna et al. 2006), is only exhibited by <0.2% of angiosperms (Baker and Whiting 2002), making the selection of native species for phytoremediation efforts a difficult task. One promising species, *Typha angustifolia,* has showed high tolerance to Cr (Dong et al. 2007).

Natural Phytoextraction

In the natural setting, certain plants have been identified, which have the potential to take heavy metals up. At least 45 plant families have been identified to have plant species that can hyperaccumulate contaminants. Some of these families are: Brassicaceae, Fabaceae, Euphorbiaceae, Asteraceae, Lamiaceae and Scrophulariaceae (Salt et al. 1998; Dushenkov 2003). Among the best-known hyperaccumulator species is *Thlaspi caerulescens,* commonly known as alpine pennycress (Kochian 1996). It is impressive that without showing injury this species accumulated up to 26,000 mg/kg Zn and up to 22% of soil exchangeable Cd from a contaminated site (Brown et al. 1995; Gerard et al. 2000). *Brassica juncea,* commonly called Indian mustard, has also been found to have the ability to transport lead from its roots to the shoots.

Results of studies performed worldwide have shown that certain plants will tolerate high levels of metals in their tissues. For example, Ni is known to reside at concentrations of >1,000 mg/kg in more than 320 plant species (spp.). Besides Ni, high concentrations of other elements have also been found in many plant species (number of species given in the parenthesis) as follows: Co (30 spp.), Cu (34 spp.), Se (20 spp.), Pb (14 spp.), and Cd (1 sp.) (Reeves and Baker 2000). Concentrations exceeding 10,000 mg/kg have been recorded for Zn (11 spp.) and Mn (10 spp.). The hyperaccumulation threshold levels of these elements have been set higher because their normal ranges in plants (20–500 mg/kg) are much higher than that for the other heavy metals (Reeves 2003). Aquatic plants such as the floating *Eichornia crassipes* (water hyacinth) (Mohanty and Patra 2007), *Lemna minor* (duckweed), and *Azolla pinnata* (water velvet) have been investigated for use in rhizofiltration, phytodegradation, and phytoextraction (Salt et al. 1997).

Induced Phytoextraction or Chelate-assisted Phytoextraction

Within the plant cell, heavy metal residues may trigger the production of oligopeptide ligands known as phytochelatins (PCs) and metallothioneins (MTs) (Ghosh and Singh 2005a). These peptides bind and form stable complexes with the heavy metal and thus neutralize the toxicity of the metal ions. PCs are synthesized with glutathione as a building block and results in a peptide with the structure: Gly-(γ-Glu-Cys-)n, {where $n = 2-11$}. Appearance of phytochelating ligands has been reported in hundreds of plant species exposed to heavy metals. MTs are small gene encoded Cys-rich polypeptides. PCs are functionally equivalent to MTs. In addition to use of PCs, one can enhance uptake effectiveness by adding the synthetic chelate EDTA to the soil (Huang et al. 1997). Similar results can be achieved by using citric acid to enhance uranium uptake. The results of Huang et al. (1997) showed that chelates enhance or facilitate Pb transport into the xylem, and increase lead translocation from roots to shoots. For the chelates tested, the order of effectiveness in increasing Pb desorption from the soil was EDTA > hydroxyethylethylene-diaminetriacetic acid (HEDTA) > diethylenetriaminepentaacetic acid (DTPA) > ethylenediamine di (o-hyroxyphenylacetic acid) (EDDHA) (Huang et al. 1997).

Limitations of Phytoextraction

Phytoextraction and plant-assisted bioremediation are most effective in the removal of contaminants if soil contamination resides within 3 ft of the surface and if ground-water is within 10 ft of the surface. It is uncertain whether an approach based on chemical chelators is practical for improving phytoextraction, since chemical chela-tors are also toxic to plants and may increase the uptake of metals but decrease plant growth.

Phytovolatilization

Phytovolatilization is another type of phytoremediation. Phytovolatilization involves the use of plants to take up contaminants from the soil, transform them into a volatile form, and then transpire them into the atmosphere. Phytovolatilization occurs as growing trees and other plants take up water and any associated organic and inorganic contaminants. Some of these contaminants can pass through plant membranes and ultimately to the leaves, where they can volatilize into the atmo-sphere at comparatively low concentrations (Fig. 5) (Mueller et al. 1999).

Phytovolatilization has been primarily used for the removal of mercury; the mercuric ion is transformed in plants into the less toxic elemental mercury. The dis-advantage of this process is that mercury released into the atmosphere is likely to be recycled by precipitation and then redeposited back into ecosystem (Henry 2000). Bioremediation of Se and As is also possible using phytovolatilization technique.

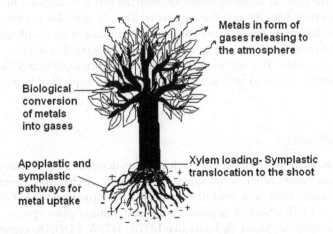

Metals in form of gases releasing to the atmosphere

Biological conversion of metals into gases

Apoplastic and symplastic pathways for metal uptake

Xylem loading- Symplastic translocation to the shoot

Fig. 5 Pictorial representation of the process of phytovolatilization

Rhizofiltration

Rhizofiltration is a technique that can remove contaminants from flowing water and aqueous waste streams by processing them through the extensive root system of plants. Several aquatic plant species and hyperaccumulator plants have been used to remove heavy metals from waste stream/water.

Rhizofiltration is defined as the use of plants, both terrestrial and aquatic, to absorb, concentrate, and precipitate contaminants from polluted aqueous sources that have a low contaminant concentration in their roots. Rhizofiltration can be used to partially treat industrial discharge, agricultural runoff, or acid-mine drainage. It can be used to remove lead, cadmium, copper, nickel, zinc, and chromium, which are primarily retained within plant roots (Chaudhury et al. 1998; USPAR 2000). The advantages of rhizofiltration include the fact that (a) it can be used in in situ or ex situ applications and (b) with non-hyperaccumulator plant species. Several plants (e.g., sunflower, Indian mustard, tobacco, rye, spinach, and corn) have been studied for their ability to remove lead from effluents by rhizofiltration. Sunflower has the greatest ability to remove lead contamination among the tested species. Wetland plant species such as water hyacinth (*Eichornia crassipes*), duckweed, smooth cord grass, smartweed, *Thlaspi caerulescens*, some members of Brassicaceae family have also been used to remove heavy metals, such as As(V), Cd(II), Cr(VI), Cu(II), Ni(II), Se(VI), Zn, and Pb.

Phytostabilization

Phytostabilization is the process in which plants are used to transform soil metals to less toxic forms, but without removing the metal from the soil (Fig. 6). It is mainly used for remediation of soil, sediment, and sludges and depends on the ability of plant roots to limit contaminant mobility and bioavailability in the soil (USPAR 2000; Mueller et al. 1999). Phytostabilization can occur through sorption, precipitation, complexation, or metal valence reduction (Salt et al. 1995). The primary purpose of the plant in fulfilling its phytostabilizing role is to decrease the amount of water that percolates through the soil matrix. This, in turn, slows or prevents the formation of hazardous leachate and prevents soil erosion and the distribution of the toxic metal to other areas. It is the dense root system of plants that stabilizes the soil and prevents erosion. This approach is very effective when rapid immobilization is needed to preserve ground and surface water, and when disposal of biomass is not required.

Phytodetoxification

This method is an in situ process, which involves detoxification of heavy metals through plant-based chelation, reduction, and oxidation mechanisms. Several plant species and algae have been used in the reduction of chromium(VI) to Cr(V), and eventually to Cr(III). Several vegetable crops and wetland plant species are also used for remediation. Metal chelators like EDTA, DTPA, EDDHA, organic acids

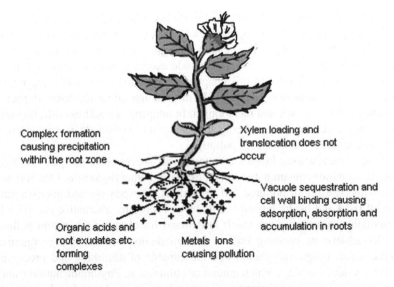

Complex formation
causing precipitation
within the root zone

Xylem loading and
translocation does not
occur

Vacuole sequestration and
cell wall binding causing
adsorption, absorption and
accumulation in roots

Organic acids and
root exudates etc.
forming
complexes

Metals ions
causing pollution

Fig. 6 Pictorial representation of the process of phytostabilization

(carboxylic acids, salicylic acids, maleic acid), and glycine are also useful for heavy metals remediation (Henry 2000; Salt et al. 1998).

6.2.3 Genetic Engineering to Improve Phytoremediation

To breed plants of higher biomass that have superior phytoremediation potential is an attractive avenue for improving phytoremediation. The high biomass and productivity of a plant is genetically controlled. Therefore introduction of such desired genes to a plant (transgenic plant) will improve the phytoremediation activity. Notwithstanding, it is crucially important to identify appropriate plant genotypes/cultivars that are resistant/tolerant to Cr, or to other contaminants, so that new and relevant genes become available for use in environmental remediation. To achieve a better understanding of the physiological and biochemical mechanisms of heavy metal tolerance/resistance in plants, researches in the related areas are essential. These research areas are fundamental for discovering or creating new metal-resistant plant species. Such work has been undertaken by Dong et al. (2007). These authors are using *Typha angustifolia* plant species that grow in Cr-contaminated areas and demonstrate high resistance to Cr stress.

7 Summary

Chromium is an important toxic environmental pollutant. Chromium pollution results largely from industrial activities, but other natural and anthropogenic sources also contribute to the problem. Plants that are exposed to environmental contamination by chromium are affected in diverse ways, including a tendency to suffer metabolic stress. The stress imposed by Cr exposure also extends to oxidative metabolic stress in plants that leads to the generation of active toxic oxygen free radicals. Such active free radicals degrade essential biomolecules and distort plant biological membranes. In this chapter, we describe sources of environmental chromium contamination, and provide information about the toxic impact of chromium on plant growth and metabolism. In addition, we address different phytoremediation processes that are being studied for use worldwide, in contaminated regions, to address and mitigate Cr pollution.

There has been a long history of attempts to successfully mitigate the toxic effects of chromium-contaminated soil on plants and other organisms. One common approach, the shifting of polluted soil to landfills, is expensive and imposes environmental risks and health hazards of its own. Therefore, alternative eco-friendly bioremediation approaches are much in demand for cleaning chromium-polluted areas. To achieve its cleaning effects, bioremediation utilizes living organisms (bacteria, algae, fungi, and plants) that are capable of absorbing and processing chromium residues in ways which amend or eliminate it. Phytoremediation (bioremediation with plants) techniques are increasingly being used to reduce heavy metal contamination and to minimize the hazards of heavy metal toxicity. To achieve this, several processes, viz., rhizofiltration, phytoextraction, phytodetoxification, phytostabilization, and phytovolatilization, have been developed and are showing utility in practice, or promise. Sources of new native hyperaccumulator plants for use at contaminated sites are needed and constitute a key goal of ongoing phytoremediation research programs. Such new plants are needed to enhance the attractiveness of phytoremediation as an effective, affordable, and eco-friendly technique to achieve successful clean-up of metal-contaminated sites worldwide.

Acknowledgments The authors are grateful to University Grants Commission, New Delhi, for financial support under the schemes of UGC-DRS -SAP-II and RFSMS-UGC-DRS.

References

Adler T (1996) Aerobic and Anaerobic biodegradation of PCBS: A review. Crit Rev Biotech 10: 241–251.

Adriano DC (1986) Trace elements in the environment. Chapter 5: Chromium. Springer, New York, NY, pp 105–123.

Agency for Toxic Substances and Disease Registry (ATSDR) (1998) Toxicological profile for chromium (update). U.S. Department of Health and Human Services. Public Health Service, Cincinnati, OH.

Agency for Toxic Substances and Disease Registry (ATSDR) (2001) Toxicological profile for chromium. U.S. Department of Health and Human Services, Public Health Service, Atlanta, GA.

Ajmal M, Nomani AA, Ahmad A (1984) Acute toxicity to electroplating wastes to microorganisms. Adsorption of chromite and chromium (VI) on a mixture of clay and sand. Water Air Soil Pollut 2: 119–127.

Alexander J, Ashet J (1995) Uptake of chromate in human red blood cells and isolated rat liver cells: The role of the anion carrier. Analyst 120: 931–933.

Ali MB, Vajpayee P, Tripathi RD, Rai UN, Singh SN, Singh SP (2003) Phytoremediation of lead, nickel and copper by Salix acmophylla Boiss.: Role of antioxidant enzymes and antioxidant substances. Bull Environ Contam Toxicol 70: 462–469.

Arslan P, Beltrame M, Tomasi A (1987) Intracellular chromium reduction. Biochem Biophys Res Commun 206: 829–834.

Athalye VV, Ramachandran V, D'Souza DJ (1995) Influence of chelating agents on plant uptake of ^{51}Cr, ^{210}Pb and ^{210}Po. Environ Pollut 89: 47–53.

Atta Aly MA, Shehata NG, Kobbia TM (1999) Effect of Cobalt on tomato plant growth and mineral content. Ann Agril Sc (Cairo) 36: 617–624.

Austenfeld FA (1979) The effect of Ni, Co and Cr on net photosynthesis of primary and secondary leaves of *Phaseolus vulgaris* L. Photosynthetica 13: 434–438.

Baker AJM, Whiting SN (2002) In search of the Holy Grail – a further step in understanding metal hyperaccumulation? New Phytologist 155: 1–4.

Banu KS, Ramaswamy K (1997) Dual inoculation of vesicular arbuscular mycorrhiza and Rhizobium in green gram. Legume Res 2(3): 177–180.

Bartlett RJ, Kimble JM (1976) Behavior of chromium in soils. II. Hexavalent forms. J Environ Qual 5(4): 383–386.

Bassi M, Donini A (1984) Phyllotoxin visualization of F-actin in normal and chromium-poisoned *Euglena* cells. Cell Biol Int Rep 8: 867–871.

Bassi M, Grazia M, Ricci A (1990) Effects of chromium(VI) on two fresh water plants, *Lemna minor* and *Pistia startiotes*. 2 Botanical and physiological observations. Cytobios 62: 101–109.

Berrow ML, Webber J (1972) Trace elements in sewage sludges. J Sci Food Agric 23: 93–100.

Bianchi V, Levis AG (1977) Recent advances in chromium genotoxicity. Toxicol Environ Chem 15: 1–24.

Bishnoi NR, Dua A, Gupta VK, Sawhney SK (1993) Effect of chromium on seed germination seedling growth and yield of peas. Agric Ecos Environ 47(1): 47–57.

Blaylock MJ, Huang JW (2000) Phytoextraction of metals. In: Raskin I, Ensley BD (eds) Phytoremediation of toxic metals: Using plants to clean-up the environment. Wiley, New York, NY, pp 53–70.

Bonet A, Poschenrieder CH, Barcelo J (1991) Chromium III- Iron interaction in Fe-deficient and Fe-sufficient bean plants. I. Growth and nutrient content. J Plant Nutr 14(4): 403–414.

Brady D, Letebele B, Duncan JR, Rose PD (1994) Bioaccumulation of metals by *Scenedesmus*, *Selenastrum* and *Chlorella* algae. Water SA 20: 213–218.

Brochiero E, Bonaly J, Mestre JC (1984) Toxic action of hexavalent chromium on *Euglena gracilis* strain Z grown under heterotrophic conditions. Arch Environ Contam Toxicol 13: 603–608.

Brooks RR, Chambers MF, Nicks LJ, Robinson BH (1998) Phytomining. Trends Plant Sci 1: 359–362.

Brown SL, Chaney RL, Angle JS, Baker AJM (1995) Zinc and cadmium uptake by hyperaccumulator Thlaspi caerulescens grown in nutrient solution. Soil Sci Soc Am J 59: 125–133.

Cary EE, Allaway WH, Olson OE (1977a) Control of Cr concentrations in food plants. I. Absorption and translocation of Cr by plants. J Agric Food Chem 25(2): 300–304.

Cary EE, Allaway WH, Olson OE (1997b) Control of chromium concentrations in food plants. II. Chemistry of chromium in soils and its availability to plants. J Agric Food Chem 25: 305–309.

Cervantes C, Garcia JC, Devars S, Corona FG, Tavera HL, Carlos Torres-guzman J, Sanchez RM (2001) Interactions of chromium with micro-organisms and plants. FEMS Microbiol Rev 25: 335–347.

Chandra P, Sinha S, Rai UN (1997) Bioremediation of Cr from water and soil by vascular aquatic plants. In: Kruger EL, Anderson TA, Coats JR (eds) Phytoremediation of soil and water contaminants. ACS Symposium Series #664. American Chemical Society, Washington, DC, pp 274–282.

Chaudhury TM, Hayes WJ, Khan AG, Khoo CS (1998) Phytoremediation – focusing on accumulator plants that remediate metalcontaminated soils. Aust J Ecotoxicol 4: 37–51.

Cheng, L, Liu S, Dixon K (1998) Analysis of repair and mutagenesis of chromium-induced DNA damage in yeast, mammalian cells, and transgenic mice. Environ Health Perspect 106: 1027–1032.

Choudhury S, Panda SK (2004) Induction of oxidative stress and ultrastructural changes in moss *Taxithelium nepalense* under lead (Pb) and Arsenic (As) phytotoxicity. Curr Sci 87(3): 342–348.

Choudhury S, Panda SK (2005) Toxic effects, oxidative stress and ultrastructural changes in moss *Taxithelium nepalense* (Schwaegr) Broth. chromium and lead phytotoxicity. Water Air Soil Pollut 167(1–4): 73–90.

Cooper EM, Sims JT, Cunningham SD, Huang JW, Berti WR (1999) Chelate-assisted phytoextraction of lead from contaminated soil. J Environ Qual 28: 1709–1719.

Corradi MG, Bianchi A (1993) Chromium toxicity in *Salvia sclarea* – I. Effects of hexavalent chromium on seed germination and seedling development. Environ Exp Bot 33(3): 405–413.

Corradi MG, Gorbi G, Ricci A, Torelli A, Bassi AM (1995) Chromium induced sexual reproduction gives rise to a Cr tolerant progeny in *Scenedesmus acutus*. Ecotoxicol Environ Saf 32: 12–18.

Costa M (1997) Toxicity and carcinogenicity of Cr(VI) in animal models and humans. Crit Rev Toxicol 27(5): 431–442.

Cunningham SD, Ow DW (1996) Promises and prospects of phytoremediation. Plant Physiol 110: 715–719.

Cunningham SD, Shann JR, Crowley DE, Anderson TA (1997) Phytoremediation of contaminated water and soil. In: Kruger EL, Anderson TA, Coats JR (eds) Phytoremediation of soil and water contaminants. ACS Symposium Series #664, American Chemical Society, Washington, DC, pp 2–19.

CzakoVer K, Batle M, Raspor P, Sipiczki M, Pesti M (1999) Hexavalent chromium uptake by sensitive and tolerant mutants of *Schizosaccharomyces pombe*. FEMS Microbiol Lett 173: 109–115.

Debatto R, Luciani S (1988) Toxic effect of chromium on cellular metabolism. Sci Total Environ 71: 365–377.

Dehn B, Schuepp H (1989) Influence of VA mycorrhizae on the uptake and distribution of heavy metals in plants. Agric Ecosyst Environ 29: 79–83.

del Val C, Barea JM, Azcon Aguilar C (1999) Assessing the tolerance to heavy metals of arbuscular mycorrhizal fungi isolated from sewage sludge-contaminated soils. Appl Soil Ecol 11: 261–269.

Dixit V, Pandey V, Shyam R (2002) Chromium ions inactivate electron transport and enhance super oxide generation invitro in Pea (*Pisum sativum* L.cv. Azad) root mitochondria. Plant Cell Environ 25: 687–693.

Dong J, Wu F, Huang R, Zang G (2007) A Chromium tolerant plant growing in Cr-contaminated land. Int J Phytoremediat 9: 167–179.

Dreyfuss J (1964) Characterization of a sulfate and thiosulfate transporting system in *Salmonella typhimurium*. J Biol Chem 239: 2292–2297.

Dubey SK, Rai LC (1987) Effect of chromium and tin on survival, growth; carbon fixation, heterocyst differentiation, nitrogenase, nitrate reductase and glutamine synthetase activities of *Anabaena doliolum*. J Plant Physiol 130: 165–172.

Dushenkov D (2003) Trends in phytoremediation of radionuclides. Plant Soil 249: 167–175.

Dvorak DH, Hedin RS, Edeborn HM, Mc Intire PE (1992) Treatment of metal-contaminated water using bacterial sulfate reduction. Results from pilot-scale reactors. Biotechnol Bioeng 40: 609–616.

Erenoglu B, Patra HK, Khodr H, Romheld V, and von Wiren N (2007) Uptake and apoplastic retention of EDTA and phytosiderophore-chelated chromium (III) in maize. J Plant Nutr Soil Sci 170: 788–795.

Fasulo MP, Bassi M, Donini A (1983) Cytotoxic effects of hexavalent chromium in Euglena gracilis. II. Physiological and ultrastructural studies. Protoplasma 114: 35–43.

Frey BE, Riedl GF, Bass AE, Small LF (1983) Sensitivity of estuarine phytoplankton to hexavalent chromium. Estuar Coast Shelf Sci 17: 181–187.

Galli A, Boccardo P, Del Carratore R, Cundari E, Bronzetti G (1985) Conditions that influence the genetic activity of potassium dichromate and chromium chloride in *Saccharomyces cerevisiae*. Mutat Res 144: 165–169.

Gardea Torresdey JL, Hernandez A, Tiemann KJ, Bibb J, Rodriguez O (1998) Adsorption of toxic metal ions from solution by inactivated cells of *Larrea tridentata* (Creosote Bush). J Hazard Sub Res 1: 3–1.

Gaur A, Adholeya A (2004) Prospect of arbuscular mycorrhizal fungi in phytoremediation of heavy metal contaminated soils. Curr Sci 86(4): 528–534.

Gaw HZ, Soong PN (1942) Nodulation and dry weight of garden peas as affected by sulphur and sulphates. J Am Soc Agron 34: 100–103.

Gerard E, Echevarria G, Sterckeman T, Morel JLP (2000) Availability of Cd to three plant species varying in accumulation pattern. J Environ Qual 29: 1117–1123.

Ghosh M, Singh SP (2005a) A review on phytoremediation of heavy metals and utilization of its by products. Appl Ecol Environ Res 3(1): 1–18.

Ghosh M, Singh SP (2005b) A comparative study of cadmium phytoextraction by accumulator and weed species. Environ Pollut 133: 365–371.

Hara T, Sonoda Y (1979) Comparison of the toxicity of heavy metals to cabbage growth. Plant Soil 51: 127–133.

Hara J, Sonada Y, Iwai I (1976) Growth response of cabbage plants to transition elements under water culture conditions. I. Titanium, vanadium, chromium, manganese and Iron. Soil Sci Plant Nutr 22: 307–315.

Hauschild MZ (1993) Putrescine (1,4-diaminobutane) as an indicator of pollution-induced stress in higher plants: Barely and rape stressed with Cr(III) or Cr(VI). Ecotoxicol Environ Saf 26: 228–247.

Henderson G (1989) A comparison of the effects of chromate, molybdate and cadmium oxide on respiration in the yeast *Saccharomyces cerevisiae*. Biol Metals 2: 83–88.

Henry JR (2000) An overview of phytoremediation of lead and mercury. National Network of Environmental Management Studies (NNEMS) Report, pp 1–31.

Huang JW, Chen J, Berti WR, Cunningham SD (1997) Phytoremediation of lead contaminated soils-role of synthetic chelates in lead phytoextraction. Environ Sci Tech 31: 800–806.

Huffman EWD, Allaway WH (1973) Chromium in pants: Distribution in tissues, organelles and extracts and availability of bean leaf chromium to animals. *J Agric Food Chem* 21: 982–986.

Hunter JG, Vergnano O (1953) Trace element toxicities in oat plants. In: Marsh RW, Thomas I (Eds) Annals of Applied Biology. University Press, Cambridge, pp 761–776.

Jamal A, Ayub N, Usman M, Khan AG (2002) Arbuscular mycorrhizal fungi enhance zinc and nickel uptake from contaminated soil by soyabean and lentil. Int J Phytoremediat 4: 205–221.

Jena AK, Mohanty M, Patra HK (2004) Phyto-remediation of environmental chromiun – A review. e-Planet 2(2): 100–103.

Kabata-Pendias A, Pendias H (1992) Trace elements in soils and plants, 2nd ed. CRC Press, London, pp 227–233.

Kadiiska MB, Xiang QH, Mason RP (1994) In-vivo free radical generation by chromium (VI): An electron resonance spin trapping investigation. Chem Res Toxicol 7: 800–805.

Kaldorf M, Kuhn AJ, Schroder WH, Hildebrandt U, Bothe H (1999) Selective element deposits in maize colonized by a heavy metal tolerance conferring arbuscular mycorrhizal fungus. J Plant Physiol 154: 718–728.

Katz SA, Salem H (1994) The biological and environmental chemistry of chromium. VCH Publishers, Inc., New York, NY, ISBN 1-56081-629-5, 214p.

Kawanishi S, Inoue S, Sano S (1986) Mechanism of DNA cleavage induced by sodium chromate (VI) in the presence of hydrogen peroxidase. J Biol Chem 261: 5952–5958.

Kessler J, Sharkey AGJ, Friedal RA (1971) Spark source mass spectrometer investigation of coal particles and coal ash. US Bureau of mines. Technical Program Report 42.

Kharab P, Singh I (1985) Genotoxic effects of potassium dichromate, sodium arsenite, cobalt chloride and lead nitrate in diploid yeast. Mutat Res 155: 117–120.

Kiling J (1997) Phytoremediation of organics moving rapidly into field trials. Environ Sci Tech 31: 129 A.

Kleiman ID, Cogliatti DH (1998) Chromium removal from aqueous solutions by different plant species. Environ Technol 19: 1127–1132.

Kochian L (1996) Mechanism of heavy metal transport across plant cell membranes In: International Phytoremediation Conference, Southborough, MA. May 8–10.

Koenig P (1910) Stuien iiberdie stimulienenden and toxischen Wirkungen der varschiednwertigen chromver bindungen out die pflanzen. Landwirtsch Jehrb 39: 775–916.

Koening P (1911) The stimulatory effects of chromium compounds in plants. Chemikerzeitung 35: 442–443.

Krishnamurthy S, Wilkens MM (1994) Environmental chemistry of Cr. Northeast Geol 16(1): 14–17.

Krupa Z, Baszynski T (1995) Some aspects of heavy metal toxicity towards photosynthetic apparatus – direct and indirect effect of light and dark reactions. Acta Physiol Plant 17: 177–190.

Kumar P, Dushenkov V, Motto H, Raskin I (1995) Phytoextraction: The use of plants to remove heavy metals from soils. Environ Sci Technol 29: 1232–1238.

Laheurte F, Leyval C, Berthelin J (1990) Root exudates of maize, pine and beech seedlings influenced by mycorrhizal and bacterial inoculation. Symbiosis 9: 111–116.

Lahouti M, Peterson PJ (1979) Chromium accumulation and distribution in crop plants. J Sci Food Agric 30: 136–142.

Leung M (2004) Bioremediation: Techniques for cleaning up a mess. Biotech J 2: 18–22. Retrieved from www.biotech.ubc.ca.

Levis AG, Buttignol M, Vettorato L (1975) Chromium cytotoxic effects on mammalian cells in vitro. Atti Assoc Genet Ital 20: 9–12.

Li XL, George E, Marschner H (1991) Phosphorus depletion and pH decrease at the root-soil and hyphae-soil interfaces of VA mycorrhizal white clover fertilized with ammonium. New Phytol 119: 397–404.

Liu DH, Jaing WS, Li MX (1993) Effect of chromium on root growth and cell division of *Allium cepa*. Isr J Plant Sci 42: 235–243.

Liu KJ, Jiang J, Shi X, Gabrys H, Walczak T, Swartz M (1995) Low-frequency EPR study of chromium (V) formation from chromium (VI) in living plants. Biochem Biophys Res Commun 206: 829–834.

Loveley DR, Widman PK, Woodward JC, Phillips JP (1993) Reduction of uranium by cytochrome c3 of *Desulfovibrio vulgaris*. Appl Environ Microbiol 59: 3572–3576.

Lyon GL, Brooks RR, Peterson PJ, Bulter GW (1970) Some trace elements in plants from serpentine soil. N Z J Sci 13: 133–139.

Lytle CM, Lytle FW, Yang N, Qian JH, Hansen D, Zayed A, Terry N (1998) Reduction of Cr(VI) to Cr(III) by wetland plants: Potential for in situ heavy metal detoxification. Environ Sci Technol 32(20): 3087–3093.

Marre E (1979) Integration of solute transport in *cereals*. In: Laidman DL and Jones RG (eds) Recent advances in the biochemistry of cereals. Academic Press, New York, NY, pp 3–25.

Marre E, Lado P, Rasin Caldogno F, Colombo R, De Michelis MI (1974) Evidence for the coupling of proton extrusion to K$^+$ ion uptake in pea internode segments treated in fusicoccin or auxin. Plant Sci Lett 3: 365–379.

Marschner H (1995) Mineral nutrition of higher plants, 2nd ed. Academic press, Harcourt Brace and Co., New York, NY. ISBN: 0124735428, 9780124735422, 889p.

McGrath SP (1982) The uptake and translocation of Tri and hexavalent chromium and effects on the growth of Oat in flowing nutrient solution and in soil. New Phytol 92: 381–390.

McGrath SP (1995) Chromium and Nickel. In: Alloway BJ (ed) Heavy metals in soil. Chapman and Hall, London, pp 139–155.

McGrath SP, Smith S (1990) Chromium and Nickel. In: Alloway J (ed) Heavy metals in soils. Wiley, New York, NY, pp 125–150.

McGrath SP, Shen ZG, Zhao FJ (1997) Heavy metal uptake and chemical changes in the rhizosphere of *Thlaspi caerulescens* and *Thlaspi ochroleucum* grown in contaminated soils. Plant Soil 188: 153–159.

McNeil KR, Waring S (1992) Vitrification of contaminated soil. In: Rees JF (ed) Contaminated land treatment technologies. Society of Chemical Industry, Elsevier Applied Sciences, London, pp 143–159.

Meagher RB (2000) Phytoremediation of toxic elemental and organic pollutants. Curr Opin Plant Biol 3: 153–162.

Mertz W (1969) Chromium occurrence and function in biological system. Physio Rev 49: 163–239.

Mishra AC (2002) An attempt on improvement of nitrogen fixation in mung bean (*Vigna radiata* L. Wilczek) grown in chromite mine area soil. Ph. D. Thesis submitted to Utkal University, Bhubaneswar, Orissa.

Mishra S, Singh V, Srivastava S, Srivastava R, Srivastava M, Das S, Satsang G, Prakash S (1995) Studies on uptake of trivalent and hexavalent Cr by maize (*Zea mays*). Food Chem Toxic 33(5): 393–397.

Misra AK, Pattnaik R, Thatoi HN, Padhi GS (1994) Study on growth and N2 fixation ability of some leguminous plant species for reclamation of mine spoilt areas of Eastern Ghats of Orissa. Final Technical Report submitted to Ministry of Environment and Forests, Govt. of India.

Misra AK, Thatoi HN, Dutta B, Pattnaik MM, Padhi GS (2004) Stabilisation and restoration of ecosystem in iron and chromite mine waste areas of Eastern Ghats of Orissa through application of microbial technology. Final Technical Report submitted to Ministry of Environment and Forests, Government of India.

Mohanty M, Patra HK (2007) Water hyacinth- A tool for Green remediation. Sabujima 15: 41–43. ISSN: 0972-8562.

Mohanty M, Patra HK (2009) Attenuation of chromium toxicity in rice by chelating agents. In: Patra HK (ed) Attenuation of stress impacts on plants. Proc. Natl. Sem UGC-DRS (SAP-II). Utkal University, Orissa, pp 53–61.

Mohanty M, Jena AK, Patra HK (2005) Effect of chelated Chromium compounds on chlorophyll content and activities of catalase and peroxidase in wheat seedlings. Indus J Agric Biochem 18(1): 25–29 ISSN: 0970-6399.

Mohanty M, Jena AK, Patra HK (2008) Application of chromium and chelating agents on growth and Cr bioaccumulation in wheat (*Triticum aestivum* L.) seedlings. J Adv Plant Sci 4(1, 2): 21–26, ISSN: 0971-9350.

Mohanty M, Pattanaik MM, Misra AK, Patra HK (2009) Attenuation of Cr(VI) from chromite mine waste water by phytoremediation technology. In: Patra HK (ed) Attenuation of stress impacts on plants. Proc. Natl. Sem. UGC-DRS (SAP-II). Utkal University, Orissa, pp 19–28.

Moral R, Pedreno JN, Gomez I, Mataix J (1993) Effects of chromium on the nutrient elements content and morphology of tomato. J Plant Nutr 18: 815–822.

Mueller B, Rock S, Gowswami D, Ensley D (1999) Phytoremediation decision tree. Prepared by – Interstate Technology and Regulatory Cooperation Work Group, Lucknow, pp 1–36.

Musgrove S (1991) An assessment of the efficiency of remedial treatment of metal polluted soil. In: Proceedings of the International Conference on Land Reclamation, University of Wales. Elsevier Science Publication, Essex.

Myttenaere C, Mousny JM (1974) The distribution of chromium (VI) in lowland rice in relation to the chemical form and to the amount of stable chromium in the nutrient solution. Plant Soil 41: 65–72.

Nayak S, Rath, SP, Patra HK (2004) The physiological and cytological effect of Cr(VI) on lentil (*Lens culinaris* Medic.) during seed germination and seeding growth. Plant Sci Res 1 and 2: 16–23.

Nayak S, Patra HK, Rath SP (2008) Biochemical and Cytological basis of toxicity lesions produced by Cr(III) in germinating seeds of Lentil (*Lens culinaris* Medic.) Asian J Microbiol Biotech Environ Sci 9(4): 1–6.

Nayari HF, Szalai T, Kadar I, Castho P (1997) Germination characteristics of pea seeds originating from a field trial treated with different levels of harmful elements. Acta Argon Hung 45: 147–154.

Negri MC, Hunchman RR (1996) Plants that remove contaminants from the environment. Lab Med 27: 36–40.

Nieboer E, Richardson DHS (1980) The replace of the nondescript term 'heavy metals' by abiologically and chemically significant classification of metal ions. Environ Pollut Ser B: 3–26.

Nieboer E, Jusys AA (1988) Biologic chemistry of Cr. In: Nriagu JO, Nieboer E (eds) Chromium in the natural and human environments. Wiley, New York, NY, pp 21–80.

Nicks L, Chambers MF (1994) Nickel farm. *Discover*. September, p. 19.

Ohtake H, Silver S (1994) Bacterial detoxification of toxic chromate. In: Chaudry GR (ed) Biological degradation and bioremediation of toxic chemicals. Dioscorides Press, Portland, pp 403–415.

Olsson PA, Francis R, Read DJ, Soderstron B (1998) Growth of arbuscular mycorrhizal mycelium in calcareous dune sand and its interaction wit other soil microorganisms as estimated by measurement of specific fatty acids. Plant Soil 201: 9–16.

Paknikar KM, Bhide JV (1993) Aerobic reduction and biosorption of chromium by a chromate resistant *Aspergillus* spp. In: Torma AE, Apel ML, Brierley CL (eds) Biohydrometallurgical technologies. The Minerals, Metals and Materials Society, Warrendale, PA, pp 237–244.

Panda SK (2003) Heavy metal phytotoxicity induces oxidative stress in *Taxithelium* spp. Curr Sci 84(5): 631–633.

Panda SK, Patra HK (1997) Physiology of chromium toxicity in plants-a review. Plant Physiol Biochem 24(1): 10–17.

Panda SK, Patra HK (1998) Attenuation of nitrate reductase activity by chromium ions in excised wheat leaves. Indian J Agric Biochem 2(2): 56–57.

Panda SK, Patra HK (2000a) Nitrate and ammonium ions effect on the chromium toxicity in developing wheat seedlings. Proc Natl Acad Sci India B 70: 75–80.

Panda SK, Patra HK (2000b) Does Cr(III) produces oxidative damage in excised wheat leaves. J Plant Biol 27(2): 105–110.

Panda S, Patra HK (2004) Attenuation of toxic chromium (VI) using chelate based phytoremediation in rice. e-planet 2(1): 72–75.

Panda SK, Choudhury S (2005) Chromium stress in plants. Braz J Plant Physiol 17(1): 95–102.

Panda SK, Mohapatra S, Patra HK (2002) Chromium toxicity and water stress stimulation effects in intact senescing leaves of green gram (*Vigna radiata* L. var. Wilckzeck K851). In: Panda SK (ed) Advances in stress physiology in plants. Scientific publishers, India, pp 129–136.

Panda SK, Choudhury I, Khan MH (2003) Heavy metal induce lipid peroxidation and affects antioxidants in wheat leaves. Biol Plant 46: 289–294.

Park D, Yun YS, Jo JH, Park JM (2005) Mechanism of hexavalent chromium removal by dead fungal biomass of *Aspergillus niger*. Water Res 39: 533–540.

Parr PD (1982) Effect of Orocol TL (A corrosion inhibitor) on germination and growth of bush beans. Publication No. 1761, Environmental Sciences Division, Oak Ridge National Laboratory, Oak Ridge, Tennessee.

Parr PD, Taylor FG (1980) Incorporation of Cr in vegetation through root uptake and foliar absorption pathways. Environ Exp Bot 20: 157–160.

Parr PD, Taylor Jr FG (1982) Germination and growth effects of hexavalent chromium in Orocol TL (a corrosion inhibitor) on *Phaseolus vulgaris*. Environ Int 7: 197–202.

Patnaik R (1995) Impact of dual inoculation of Rhizobium and VAM fungi on growth and nitrogen fixation of selected crop legume grown in mine area soil. Ph.D. Thesis submitted to Utkal University, Bhubaneswar, Orissa.

Patra HK, Sayed S, Sahoo BN (2002) Toxicological aspects of Cr(VI) induced catalase, perox-idase and nitrate reductase activities in wheat seedlings under different nitrogen nutritional environment. Pollut Res 21(3): 277–283.

Patra HK, Jena AK, Lenka S, Mohanty M (2005) Effect of ionic and chelated chromium complexes on mung bean seedlings during early phases of plant growth. Plant Sci Res 27(1 and 2): 66–70. ISSN: 0972-8564.

Pawlisz AV (1997) Canadian water quality guidelines for Cr. Environ Toxicol Water Qual 12(2): 123–161.

Peterson PJ (1975) Element accumulation by plants and their tolerance of toxic mineral soils. In: Hutchinson TC (ed) Proceedings of the International Conference on Heavy Metals in the Environment, vol. 2. University of Toronto, Toronto, pp. 39–54.

Pillichshammer M, Pumpel T, Poder R, Eller K, Klima A, Schinner F (1995) Biosorption of chromium to fungi. Biometal 8(2): 117–121.

Poschenrieder CH, Vazquer MD, Bonet A, Barcelo J (1991) Chromium III – Iron interaction in iron sufficient and iron deficient bean plants. II. Ultrastructural aspects. J Plant Nutr 14(4): 415–428.

Pratt PF (1966) Chromium. In: Chapman HD (ed) Diagnostic criteria for plants and soils, Chapter 9. University of California, Riverside, pp 136–141.

Quagraine EK, Peterson HG, Headley JV (2005) In situ bioremediation of naphthenic acids con-taminated tailing pond waters in the Athabasca oil sands region—demonstrated field studies and plausible options: A review. J Environ Sci Heal 40: 685–722.

Rajakaruna N, Tompkins KM, Pavicevic PG (2006) Phytoremediation: an affordable green tech-nology for the clean-up of metal-contaminated sites in Sri Lanka. Cey J Sci (Bio Sci) 35(1): 25–39.

Ramanathan S, Ensor M, Daunert S (1997) Bacterial biosensors for monitoring toxic metals. Trends Biotechnol 15: 501–506.

Ramirez R, Calvo Mendez C, Avila-Rodriguez M, Gutierrez-Corona JF (2000) Chromate resis-tance and reduction in a yeast strain isolated from industrial waste discharges. In: Raynal JA, Nucklos JR, Reyes P, Ward M (eds) Environmental engineering and health sciences, Section 4: Environmental engineering application. Water Resources Publications, LCC, Englewood, CO, pp 437–445.

Rapoport AI, Muter OA (1995) Biosorption of hexavalent chromium by yeast. Process Biochem 30: 145–149.

Raskin I, Ensley BD (2000) Phytoremediation of toxic metals: Using plants to clean up the environment. Wiley, New York, NY, pp 53–70.

Reed D, Tasker IR, Cunnane JC, Vandegrift GF (1992) Environmental restoration and separa-tion science. In: Vandgrift GF, Reed DT, Tasker IR (eds) Environmental remediation removing organic and metal ion pollutants. ACS Symposium Series 509, American Chemical Society, Washington, DC, pp 1–21.

Reeves RD (2003) Tropical hyperaccumulators of metals and their potential for phytoextraction. Plant Soil 249: 57–65.

Reeves RD, Baker AJM (2000) Metal accumulating plants. In: Raskin I, Ensley B (eds) Phytoremediation of toxic metals: Using plants to clean up the environment. Wiley, New York, NY, pp 193–229.

Rilling MC, Steinberg PD (2002) Glomalin production by an arbuscular mycorrhizal fungus: A mechanism of habitat modification. Soil Biol Biochem 34: 1371–1374.

Ross DS, Sjogren RE, Bartlett RJ (1981) Behavior of chromium in soils IV. Toxicity to microorganisims. J Environ Qual 10: 145–148.

Rulkens WH, Tichy R, Grotenhuis JTC (1998) Remediation of polluted soil and sediment: Perspectives and failures. Water Sci Technol 37: 27–35.

Salt DE, Smith RD, Raskin I (1998) Phytoremediation. Annu Rev Plant Physiol Plant Mol Biol 49: 643–668.

Salt DE, Blaylock M, Kumar NPBA, Dushenkov V, Ensley BD, Chet I, Raskin I (1995) Phytoremediation: A novel strategy for the removal of toxic metals from the environment using plants. Biotechnology 13: 468–474.

Salt DE, Pickering IJ, Prince RC, Gleba D, Dushenkov S, Smith RD, Raskin I (1997) Metal accumulation by aquacultured seedlings of Indian Mustard. Environ Sci Technol 31(6): 1636–1644.

Sampanpanish, P, Pongsapich W, Khaodhiar S, Khan E (2006) Chromium removal from soil by phytoremediation with weed plant species in Thailand. Water Air Soil Pollut Focus 6: 191–206.

Sampedro MA, Blanco A, Liama MJ, Serra JL (1995) Sorption of heavy metals to *Phormidium laminosum* biomass. Biotechnol Appl Biochem 22: 355–366.

Saner G (1980) Chromium in nutrition and disease. Curr Top Nutr Dis 2. Alan R. Liss, New York, ISBN: 08-451-16010,135p.

Sarkar A, Jana S (1987) Effect of combinations of heavy metals on hill activity of *Azolla pinnata*. Water Air Soil Pollut 35(1/2): 141–145.

Schmidt W (1996) Influence of Cr(III) on root associated Fe(III)-reductase in *Plantago lanceolata* L. J Exp Bot 47: 805–810.

Shanker, AK, Cervantes C, Loza-Tavera H, Avudainayagam S (2005a) Chromium toxicity in plants. Environ Int 31: 739–753.

Shanker AK, Ravichandran V, Pathmanabhan G (2005b) Phytoaccumulation of chromium by some multipurpose-tree seedlings. Agroforestry Syst 64: 83–87.

Shewry PR, Peterson PJ (1974) The uptake and transport of chromium by barley seedlings (*Hordeum vulgare* L.). J Exp Bot 25: 785–797.

Shi X, Dalal NS (1989) Chromium (V) and hydroxyl radical formation during the glutathione reductase – catalyzed reduction of chromium (VI). Biochem Biophys Res 163: 627–634.

Silver S, Williams JW (1984) Bacterial resistance and purification of heavy metals. Enzyme Microb Technol 6: 531–537.

Skeffington RA, Shewry PR, Peterson PJ (1976) Chromium uptake and transport in barley seedlings (*Hordeum vulgare* L.). Planta 132: 209–214.

Smith B (1993) Remediation update funding the remedy. Waste Manage Environ 4: 24–30.

Solomonson IP, Barber MJ (1990) Assimilatory nitrate reductase: Functional properties and regulation. Annu Rev Plant Physiol Plant mol Biol 41: 225–253.

Srivastava S, Prakash S, Srivastava MM (1999) Chromium mobilization and plant availability – the impact of organic complexing ligands. Plant Soil 212: 203–208.

Stern RM (1982) Chromium compounds: Production and occupational exposure. In: Langard S (ed) Biological and environmental aspects of chromium. Elsevier Biomedical Press, Amsterdam, New York, NY, pp 16–47.

Strile M, Kolar J, Selih VS, Kocar D, Pihlar B (2003) A comparative study of several transition metals in fenton like reaction system at circum-neutral pH. Acta Chin Slov 50: 619–632.

Thatoi HN (1994) Study on growth and N₂ fixation in selected tree legumes under Rhizobium and VAM fungi inoculation in Iron and chromite mine waste soil. Ph.D. Thesis submitted to Utkal University, Bhubaneswar, Orissa.

Torresdey JLG, Videa JRP, Montes M, Rosa G, Diaz CB (2004) Bioaccumulation of cadmium, chromium and copper by *Convolvulus arvensis*: Impact on plant growth and uptake of nutritional elements. *Bioresour Technol* 92: 229–235.

Torresdey JLG, Rosa G, Videa J.P, Montes M, Jimenez GC, Aguilera IC (2005) Differential uptake and transport of trivalent and hexavalent chromium by tumbleweed (*Salsola kali*). Arch Environ Contam Toxicol 48: 225–232.

Travieso L, Canizarez RO, Borja R, Benitez F, Dominguez AR, Dupeyron R, Valiente V (1999) Heavy metal removal by microalgae. Bull Environ Contam Toxicol 62: 144–151.

Tripathi RD, Chandra P (1991) Chromium uptake by *Spirodela polyrhiza* (L.) Schleiden. In: Relation to metal chelators and pH. NBRI Research Publication 367 (N.S.), pp 764–769.

Turnau K, Kottke I, Oberwinkler F (1993) Element localization in mycorrhizal roots of *Pteridium aquilinum* L. Kuhn collected from experimental plots treated with cadmium dist. New Phytol 123: 313–324.

U.S. Department of Health and Human Services (USDHHS) (1993) Registry of toxic effects of chemical substances (RTECS, online database). National Toxicology Information Program, National Library of Medicine, Bethesda, MD.

U.S. Environmental Protection Agency (USEPA) (1983) Health assessment for chromium. EPA-600/8-83-014F. Final report. Washington DC.

USEPA (1998) Toxicological review of hexavalent chromium. Support of summary information on the integrated risk information system. Washington DC, USA.

U.S. Environmental Protection Agency (USEPA) (1999) Integrated risk information system (IRIS) on Chromium III. National Center for Environmental Assessment, Office of Research and Development. Washington, DC.

U.S. Environmental Protection Agency Reports (USPAR) (2000) Introduction to phytoremediation. EPA 600/R-99/107. National Risk Management Research Laboratory, Cincinnati, OH; http://www.epa.gov/swertio1/download/remed/introphyto.pdf.

Vajpayee P, Sharma SC, Tripathi RD, Rai UN, Yunus M (1999) Bioaccumulation of chromium and toxicity to photosynthetic pigments, nitrate reductase activity and protein content of *Nelumbo nucifera* Gaertn. Chemosphere 39: 2159–2169.

Vajpayee P, Tripathi RD, Rai UN, Ali MB, Singh SN (2000) Chromium (VI) accumulation reduces chlorophyll biosynthesis, nitrate reductase activity and protein content in Nymphaea alba L. Chemosphere 41: 1075–82.

Vajpayee P, Rai UN, Ali MB, Tripati RD, Yadav V, Sinha S, Singh SN (2001) Chromium induced physiologic changes in Valisneria spiralis L. and its role in phytoremediation of tannery effluents. Bull Environ Cont Toxicol 67: 246–256.

Van Assche F, Clijsters H (1990) Effects of metals on enzyme activity in plants. Plant Cell Environ 13: 195–206.

Vazquez D, Poschenrieder CH, Barcelo J (1987) Chromium VI induced structural and ultra structural changes in bush bean plants (*Phaseolus vulgaris* L.). Ann Bot 59: 427–438.

Vivas A, Marulanda A, Gomez M, Barea JM, Azcon R (2003) Physiological characteristics (SDH and ALP activities) of arbuscular mycorrhizal colonization as affected by *Bacillus thuringiensis* inoculation under two phosphorus levels. Soil Biol Biochem 35: 987–996.

Volesky B, Holan ZR (1995) Biosorption of heavy metals. Biotechnol Prog 11: 235–250.

Vymazal J (1990) Uptake of lead, chromium, cadmium and cobalt by *Cladophora glomerata*. Bull Environ Contam Toxicol 44: 468–472.

Wakatasuki T (1995) Metal oxido-reduction by microbial cells. J Ind Microbiol 14: 169–177.

Wales DS, Sagar BF (1990) Recovery of metal ions by microfungal filters. J Chem Technol Biotechnol 49: 345–355.

Wallace A, Soufi SM, Cha JW, Romney EM (1976) Some effects of chromium toxicity on bush bean plants grown in soil. Plant Soil 44: 471–473.

Watanabe H (1984) Accumulation of chromium from fertilizer in cultivated soils. Soil Sci Plant Nutr 4: 543–554.

Wiegand HJ, Ottenwalder H, Bolt HM (1985) Determination of chromium in human red blood cells. Basis for a concept of biological monitoring. Arbeitsmed Sozialmed Praeventivmed 20: 1–4 (in German).

Williamson A, Johnson MS (1981) Reclamation of metalliferous mine wastes. In: Lepp NW (ed) Effect of heavy metal pollution on plants, vol. 2. Applied Science Publishers, Barking, Essex, pp 185–212.

Wong PT, Trevors JT (1988) Chromium toxicity to algae and bacteria. In: Nriagu JO, Nieboer E (eds) Chromium in the natural and human environments. Wiley, New York, NY, pp 305–315.

Wood JM, Wang HK (1983) Microbial resistance to heavy metals. Environ Sci Technol 17: 582–590.

World Health Organization (1988) Chromium. Environ Health Criteria 61: 197.

Yang X, Baligar VC, Martens DC, Cleark RB (1996) Plant tolerance to nickel toxicity. I. Influx, transport and accumulation of nickel in four species. J Plant Nutr 19: 73–85.

Zaccheo P, Genevini PL, Cocucci S (1982) Chromium ions toxicity on the membrane transport mechanism in segments of maize seedling roots. J Plant Nutr 5: 1217–27.

Zayed AM, Terry N (2003) Chromium in the Environment: Factor affecting biological remediation. Plant and Soil 249: 139–156.

Zayed A, Lytle CM, Qian JH Terry N (1998) Chromium accumulation, translocation and chemical speciation in vegetable crops. Planta 206: 293–299.

Zeid IM (2001) Responses of *Phaseolus vulgaris* to chromium and cobalt treatment. Biol Plant 44: 111–115.

Zhu YL, Zayed AM, QuH, De Souza M, Terry N (1999) Phytoaccumulation of trace elements by wet land plants. II. Water Hyacinth. J Environ Qual 28: 339–344.

Zhuang P, Yang QW, Wang HB, Shu WS (2007) Phytoextraction of heavy metals by eight plant species in the field. Water Air Soil Pollut 84: 235–242.

The Effects of Radionuclides on Animal Behavior

Beatrice Gagnaire, Christelle Adam-Guillermin, Alexandre Bouron, and Philippe Lestaevel

Contents

1   Introduction   36
2   Link Between Alteration of Physiological Mechanisms and Behavioral Effects   . . .   37
   2.1   Disruption of Sensorial Activity   38
   2.2   Neurological Dysfunction   39
   2.3   Endocrine Disruption   40
   2.4   Oxidative Disruption   41
   2.5   Metabolic Disruption   41
3   Behavioral Responses in Organisms   42
   3.1   Humans   42
   3.2   Rodents   44
   3.3   Fish and Wildlife Species   45
4   Consequences   49
   4.1   Consequences in Humans: Neurodegenerative Pathologies   49
   4.2   Ecological Consequences   49

B. Gagnaire (✉)
Laboratoire de Radioécologie et d'Ecotoxicologie, IRSN (Institut de Radioprotection et de Sûreté Nucléaire), Centre de Cadarache, Bat 186, 13115 Saint-Paul-Lez-Durance Cedex, France
e-mail: beatrice.gagnaire@irsn.fr

C. Adam-Guillermin
Laboratoire de Radioécologie et d'Ecotoxicologie, IRSN (Institut de Radioprotection et de Sûreté Nucléaire), Centre de Cadarache, 13115 Saint-Paul-Lez-Durance Cedex, France
e-mail: christelle.adam-guillermin@irsn.fr

A. Bouron
CEA, Institut de Recherche en Technologies et Sciences pour le Vivant, Grenoble, France, CNRS, UMR 5249, Laboratoire de Chimie et Biologie des Métaux, Grenoble, France Université Joseph Fourier, Grenoble, France
e-mail: alexandre.bouron@cea.fr

P. Lestaevel
Laboratoire de Radiotoxicologie Expérimentale, IRSN, BP17, 92262 Fontenay-aux-Roses Cedex, France
e-mail: philippe.lestaevel@irsn.fr

D.M. Whitacre (ed.), *Reviews of Environmental Contamination and Toxicology*,
Reviews of Environmental Contamination and Toxicology 210,
DOI 10.1007/978-1-4419-7615-4_2, © Springer Science+Business Media, LLC 2011

5   Conclusion and Future Work   50
6   Summary   50
References   51

1 Introduction

Behavior refers to the observable or measurable actions or reactions of an organism (movements, physiological alterations, verbal expression, etc.) in response to a stimulus originating from its environment (Bone and Moore 2008). Animals express several types of behavior including sexual, reproductive, social (aggression, maternal relationship, etc.), activity (locomotion, feeding, defence and avoidance responses) and cognitive behaviors (attention, learning, memory) (Zala and Penn 2004).

Behavior is regulated by several body systems. These include the sensorial system (chemoreception), nervous system (production of neurotransmitters including acetylcholine), endocrine system and oxidative and metabolic activities (oxygen consumption, lipids, glycogen). Even subtle disturbances in these systems may translate into serious behavioral aberrations, which can have severe implications for survival (Baatrup 2009). Behavior represents the culmination of all the anatomical adaptations and physiological processes that occur within an organism. Behavioral adaptations, along with morphological and physiological adaptations, may occur that increase the fitness of individuals and, through natural selection, the evolution of the species (Bone and Moore 2008).

Behavioral data are difficult to obtain and are often criticized because of their high variability. In addition, laboratory experiments often lack natural ecological realism, which can result in artefacts that make measuring behavior in the field difficult (Zala and Penn 2004). However, standardization of methods is improving, and behavioral testing methods have the advantage of being non-destructive and inexpensive. Moreover, most behavioral measurements induce minimal stress in test organisms and may be repeated on the same individuals, which suggests that behavioral assays could be more powerful than other methods (Kavlock et al. 1996; Little and Finger 1990).

For decades, direct mortality has been a primary metric for assessing the effects of chemical contamination of ecosystems. Ecotoxicologists were among the first to recognize that behavioral measures may have value in the study of sublethal effects of a pollutant, because they may have high sensitivity to changes in the steady state of an organism, compared to other endpoints (Peakall 1996). Moreover, behavioral changes have great potential as biomarkers of internal and external stress in animals, because behavior represents both the physical manifestation of the animal's internal neuronal, metabolic and endocrine processes, and the integrated physiological response to its environment (Clotfelter et al. 2004). Therefore, behavioral measures could be more relevant than ones based on biochemical or physiological parameters (Zala and Penn 2004).

Although behavioral studies conjoined with measures of exposure to contaminants are increasing, few studies have used behavior as an endpoint following exposures to radioactive contaminants. This is unfortunate, because the nuclear industry is expanding worldwide. Several nuclear applications, including nuclear-based energy production, medicine and research, tend to increase the environmental concentrations of some radioactive metals, such as uranium, caesium, cadmium and cobalt (IUPAC 2004). These radioactive metals have many uses in the nuclear industry. Uranium is used in the nuclear fuel cycle as an important source of energy derived from its spontaneous radioactive disintegrations and its fissile properties. Cadmium is used in the form of control rods that regulate neutron flux in reactors. Radioactive cadmium may be formed from unavoidable neutron activation of stable cadmium, as well as from the activation of other stable metals in the core. Cobalt is an essential element present in organisms as vitamin B_{12}. Its radioisotopes, ^{60}Co and ^{58}Co, are among the main activation products found in the liquid wastes of nuclear power plants, while ^{60}Co is largely used in medicine and industry (gammagraphy, sterilization). Caesium is produced by the fission of uranium in nuclear reactors and is mainly used in radiotherapy. Moreover, the radioisotopes ^{134}Cs and ^{137}Cs are the principal radionuclides measured from fallout of nuclear weapon tests and some nuclear accidents, most notably the Chernobyl disaster.

There are studies in the literature that have addressed the effects of these foregoing elements on behavior in the context of epidemiological surveys of exposed human populations or in laboratory experiments, primarily using rodents, fish, birds and invertebrates as biological models. Therefore, the purpose of this review is to provide an overview of the current literature that focuses on the chemo- and radiotoxicological effects of uranium, caesium, cadmium and cobalt on behavior; the review includes the effects of these elements on the potential underlying mechanisms of behavioral alterations and the behavioral parameters measured on the studied organisms.

2 Link Between Alteration of Physiological Mechanisms and Behavioral Effects

Behavior results from interactions between organisms and their external environment and represents the integration of physiological processes and of mechanisms happening at the subcellular level. A conceptualization of events that form the mechanistic basis of behavior is presented in Fig. 1.

When the environment is modified, any of the processes depicted in Fig. 1 may be affected in a sequential manner, from the disruption of systems receiving the information (e.g., olfactory information) to downstream neurochemical and hormonal mechanisms. An understanding of the mechanisms that control organismal behavior relies on the study of neurochemical and hormonal processes and of molecules involved in the regulation of neuronal activity, including neurotransmitters.

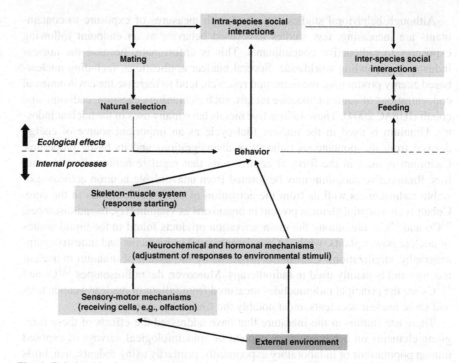

Fig. 1 Relationships between processes that control behavior and ecological effects that may result from behavioral disruption (adapted from Grue et al. 2002)

In this section, we address the description of potential mechanisms underlying behavioral alterations. Behavior is a product of the integration of many physiological systems, namely the sensory, hormonal, neurological, oxidative and metabolic systems. In this review, we address and focus on such system effects in the context of the elements used by the nuclear industry (uranium, caesium, cadmium and cobalt). Broader and exhaustive reviews of this literature on metals in general have been described elsewhere (Baatrup 1991; Clotfelter et al. 2004; Nordberg et al. 2007; Peakall 1996; Scott and Sloman 2004; Zala and Penn 2004; Zalups and Koropatnick 2000).

2.1 Disruption of Sensorial Activity

In fish, chemoreception is very well developed and plays an important role in their response to the environment. Fish rely on olfaction in predator avoidance, reaction to alarm cues and reproduction (Chivers and Smith 1998; Døving and Lastein 2009). In several studies, the effects of non-radioactive metals (e.g., cadmium, copper, zinc, nickel, manganese) have been examined on fish olfaction. The olfactory system is comprised of an olfactory epithelium, the rosette, which contains several ciliated receptor cells linked to the olfactory bulb by the olfactory nerve (Collin 2007).

Receptor cells are bipolar neurons that are in direct contact with the environment and can therefore serve as a portal for metals to enter the brain. In fish, exposure to waterborne cadmium at environmentally relevant levels caused the deposition of cadmium within olfactory sensory neurons (Blechinger et al. 2007; Scott et al. 2003). In rats, cadmium is preferentially accumulated in the brain, since this metal is transported along olfactory neurons by axonal transport mechanisms (Tallkvist et al. 2002).

In Colorado pikeminnow, *Ptychocheilus lucius*, copper accumulation in primary neurones led to the degeneration of ciliated olfactory receptor cells, but this phenomenon was reversible when fishes were placed in an uncontaminated environment (Beyers and Farmer 2001). This transitory capacity of sacrificing peripheral olfactory receptors could give a selective advantage to these fish species. However, reversibility is not always observed. A brief cadmium exposure in zebrafish larvae resulted in long-term deficits in olfaction that resulted in a reduction in response to alarm cues. These response reductions even occurred after a depuration period (Blechinger et al. 2007), probably from necrosis of the olfactory epithelium cells and alterations in the ciliated sensory cells of the olfactory pit (Matz and Krone 2007).

Other sensorial systems may also be disrupted by metals, e.g., low concentrations of cadmium induced histological alterations of the trunk lateral neuromasts in the sea bass, *Dicentrarchus labrax* (Faucher et al. 2006).

The effects of radionuclides on olfaction of fish or other vertebrates have been infrequently studied. However, a disruption of the expression of genes (*or111-7* and *or-102-5*) coding for olfactory receptors was shown in the zebrafish, *Danio rerio*, exposed to depleted uranium, together with an alteration of the ultrastructure of the olfactory bulb (Lerebours et al. 2010).

In rats, uranium has been proposed to enter the brain, not only by the common systemic routes of exposure, but also by a specific inhalation exposure route, the olfactory pathway. Hence, the inhalation of depleted uranium particles in rats led to uranium accumulation in the olfactory bulb and brain, and indicated the existence of a potential transfer pathway to the brain (Monleau et al. 2005). This accumulation of uranium affected locomotor and memory functions. Tournier et al. (2009) confirmed the existence of this olfactory pathway that led to depleted uranium accumulation, and their results showed a specific favoured frontal brain accumulation of the inhaled uranium in rats, mainly in the olfactory bulbs and tubercles, frontal cortex and hypothalamus.

2.2 Neurological Dysfunction

Cholinergic transmission plays an important role in neurocognitive functions and muscular contraction mechanisms and is affected in senile dementia pathologies. Cholinergic neurotransmission involves acetylcholine (ACh) synthesis and release. Once exocytosed, ACh can be hydrolysed by acetylcholinesterase (AChE) on post-synaptic membranes. AChE activity is a commonly used biomarker to assess altered neural functions in brain.

Brain neurotransmitter levels and enzymatic functions have been shown to correlate with behavior. Several studies on pesticides, including organophosphates and carbamates, displayed relationships between the inhibition of AChE and behavioral alterations in fishes, including rainbow trout, *Oncorhynchus mykiss* (Beauvais et al. 2000, 2001; Brewer et al. 2001), and zebrafish (Linney et al. 2004; Roex et al. 2003).

Data on similar effects caused by radionuclides are more scarce. The neurotoxic effects of uranium, particularly its implications for neurotransmission, have, however, been studied. In uranium-exposed zebrafish, a decrease of brain AChE was first observed after 36 h exposure, followed by an increase of the activity at 5–20 days (Barillet et al. 2007). Effects were identical for depleted and enriched uranium, leading to the conclusion that the chemotoxicity of this element is more important than its radiotoxicity. The consequences of AChE inhibition have been demonstrated using *ache* zebrafish mutants that possess no AChE activity: effects are a decrease in mobility, rapidly leading to total immobility (Cousin et al. 2005). The phenotype of these uncontaminated mutants was modified at the behavioral and structural levels, with a muscular disorganization capable of evoking a myopathy. Such a disorganization of muscular fibres was also observed in wild zebrafish exposed to uranium (Barillet et al. 2007), which could be a consequence of AChE inhibition observed in short-term responses. Such alterations of muscular ultrastructure may lead to effects on fish swimming behavior.

In rats, daily intramuscular injection with depleted uranium for 7 days increased AChE activity in the cortex and was associated with sensory-motor alterations (Abou-Donia et al. 2002). In contrast, AChE levels were decreased only in the cortex of rat brain after 9 months of exposure to 40 mg/L of depleted uranium in drinking water (Bensoussan et al. 2009). These results showed that the cholinergic system is affected by the chemotoxicity of uranium.

Neurotransmitters, other than ACh (indirectly measured by AChE), have important roles in neurocognitive functions. In rats, chronic exposure to depleted uranium contamination disrupted the metabolism of dopamine, an essential neurotransmitter that controls locomotion (Bussy et al. 2006). Similarly, the turn-over of serotonin, a neurotransmitter involved in the sleeping-wake cycle, anxiety and depression, was disrupted by a 9-month exposure of rats to depleted uranium contamination (Bussy et al. 2006).

Non-radioactive metals may also have multiple targets in an organism: cadmium exposure of the freshwater mussel, *Anodonta cygnea*, caused changes in filtering behavior, decreased the serotonin and dopamine levels in the central nervous system (Salanki and Hiripi 1990) and also decreased the brain AChE level (Salanki et al. 1993).

2.3 Endocrine Disruption

In some studies, a relationship between hormone levels and behavior was demonstrated, indicating that a deregulation of the endocrine system could lead to behavioral effects (reviewed in Scott and Sloman 2004). Also available are data for metals on the disruption of the hypothalamo-pituitary-interrenal (HPI) axis that controls the cortisol level, a hormone involved in the stress response. Cadmium exposure of juvenile trout induced an inhibition of the cortisol increase that normally occurred during predation, leading to an inhibition of antipredatory behavior (Scott et al. 2003). Metals also disrupt other hormones, such as growth or thyroid hormones, which play an important role in the migratory behavior in fishes (Comeau et al. 2001). Sexual hormones may also be affected, as was demonstrated for cadmium, which induced a decrease in the transcriptional activity of the estradiol receptor in trout (Vetillard and Bailhache 2005). However, no data are available for radionuclides.

2.4 Oxidative Disruption

Behavioral alterations can cause complex molecular and cellular changes. The brain is highly sensitive to oxidative stress because its antioxidant defences are poorly developed; paradoxically, neuronal tissue is rich in polyunsaturated fatty acids, the main target of lipid peroxidation. In some studies, the role manifested by oxidative stress on behavioral alterations has been observed after metal exposure. Cadmium induced oxidative stress in the central nervous system and induced locomotor dysfunctions; treatment with the antioxidant vitamin E improved these dysfunctions (Ali et al. 1993). After depleted uranium exposure, an increase in lipid peroxidation in rat brain was observed, suggesting damage had occurred to the cell membrane by induction of oxidative stress and production of free radicals (Briner and Murray 2005). Metals and radionuclides, by inducing oxidative stress, can accelerate aging and the development of several diseases (notably neurodegenerative diseases) that are linked to free-radical production (Migliore and Coppedè 2009).

2.5 Metabolic Disruption

Exposure to toxicants can disrupt various aspects of metabolism in fish, from individual (metabolic rate, swimming activity) to tissue responses (substrate availability, enzyme activity), and any of these may translate into a behavioral response (Scott and Sloman 2004).

Fish, like rainbow trout, accumulate metals (cobalt and cadmium) in their gills (Richards et al. 2001); this is likely to alter respiration and therefore behavior. Attempts have been made to relate physiological dysfunctions to behavioral state that result from metal exposure. In rainbow trout exposed to cadmium, appetite decreased and Na^+/K^+ balance was disrupted, but neither O_2 consumption rate nor swimming speed was affected, unlike copper, which decreased both parameters (McGeer et al. 2000).

Several pollutants have been shown to alter levels of metabolic substrates. Cobalt decreased the basal plasma levels of glucose in *Cyprinus carpio*, probably by reducing the rate of gluconeogenesis (Hertz et al. 1989). Cadmium inhibited the cortisol response of rainbow trout to an alarm substance; the authors hypothesized that glucose synthesis would be impaired as a result (Scott et al. 2003). Cobalt decreased muscle glycogen content and increased lactic acid levels in the blood of the tropical teleost, *Colisa fasciatus*, probably from an increase of catecholamine secretion (Nath and Kumar 1988). A subsequent decrease in available energy could have consequences on fish behavior.

The specific mechanisms by which pollutants alter metabolic substrate availability are unclear, but the observed effects probably reflect actions on enzymes that participate in metabolic activity and/or in protein metabolism (reviewed in Scott and Sloman 2004).

3 Behavioral Responses in Organisms

In this section, we review the known effects of major pollutants (uranium, caesium, cadmium and cobalt) on the behavior of humans, rodents, fish and wildlife species that encounter nuclear applications or nuclear products. Because information on the radioactive forms of these elements is not always available, particularly for cadmium and cobalt, we also cite literature references that pertain to the effects of non-radioactive forms of these elements.

3.1 Humans

3.1.1 Uranium

An extra-pyramidal syndrome (dysfunction of motricity control) that evolves over several years has been described in humans; symptoms of the syndrome include ataxia (walking trouble), nystagmus (involuntary oscillation movement of globe eye) and peripheral neuropathy. In one case that lacked any etiologic causes, these symptoms occurred during the first 3 years of the disease and were attributed to the regular hand manipulation of an uranium rod (Goasguen et al. 1982). Howland (1949) reported cases of persons accidentally contaminated with uranium, who showed behavioral problems the week after the accident; however, the behavioral responses could be attributed to fear alone. The individuals exposed to the contamination showed nervousness, unusual hyperactivity and apprehension, with exaggeration of facial and verbal expressions, which sometimes extended to incoherence. An epidemiologic study on uranium miners in Bohemia disclosed a significant increase of homicides and mental disorders compared to the general population (Tomasek et al. 1994). However, the authors of this study took into account neither the psychogenic factors linked to working conditions, nor the implications of multi-pollution resulting from these miners being also exposed to high concentrations of radon and arsenic.

During the Gulf War, there were reports of soldiers being injured by depleted uranium fragments, which sometimes became permanently embedded in their bodies. Such fragments may represent a source of chronic uranium contamination and potential injury. Injured veterans display increased uranium concentrations in excreted urine, but this occurs without nephrotoxic symptomatology. However, a neurocognitive examination showed a statistical relationship between uranium concentration in urine and decreased performance on tests designed to evaluate neurocognition. The authors suggest that at this contamination level, the kidney is not the critical target organ, and rather that the neurological or reproductive systems could be the firsts to be disrupted (McDiarmid et al. 2000).

All of the effects described in this section are primarily attributed to the chemotoxicity of uranium.

3.1.2 Caesium

Several studies performed in the years following the Chernobyl accident, in 1986, showed neurological and psychological illnesses associated with radioactive caesium (^{137}Cs) exposure in clean-up personnel ("liquidators") and people living in the contaminated areas. Among the Kirghizstan liquidators, the most significant observed effects were increased nervous system diseases and mental problems. Between 1989 and 1994, the highest cause of mortality among this liquidator population was suicide (23.2%) (Kamarli and Abdulina 1996). Moreover, a significant relationship was demonstrated for liquidators between the radiation dose levels to which they were exposed and mental disorders (Ivanov et al. 2000).

3.1.3 Non-radioactive Cadmium

At the cellular level, the effects of cadmium are similar to cobalt, with both being powerful inhibitors of several ion channels in neurons and in glial cells. However, although the *in vitro* action of cadmium on neuronal physiology has been well described, it has been demonstrated in a few studies that cadmium exposure could also alter central nervous system function. Because the half-life of cadmium's residence in humans is 15–20 years, its concentration in hair is a useful criterion to establish intoxication level. Studies of exposed children showed that high cadmium concentrations in hair are associated with learning difficulties or mental retardation (Jiang et al. 1990). In rodents (Section 2.1), cadmium is known to penetrate the brain via the nasal route through olfactory neurons. Such contamination of the olfactory epithelium probably explains why the sense of smell is altered (anosomy) among cadmium–nickel battery-factory workers (Sulkowski et al. 2000). Cadmium also slows psychomotor functions and decreases the capacity of concentration among chronically exposed individuals (Hart et al. 1989).

3.1.4 Non-radioactive Cobalt

Cobalt is the constitutive element of B_{12} vitamin (cobalamine) and plays an important role in nervous system function. Exposure to cobalt-contaminated particles by inhalation may result in neurotoxicity. Such atmospheric pollutants can directly enter the brain via olfactory neurons, and cobalt is known to accumulate in olfactory bulbs. Moreover, memory deficits were observed in people exposed to cobalt particles. In one study, the pineal gland, or epiphysis, contained 1.43 times more cobalt than did other brain structures (Jordan et al. 1997). However, measurements of trace elements in serum of aged patients showed that Alzheimer patients had significantly lower cobalt concentrations compared to age-matched subjects, suggesting that cobalt could actually act to protect some cognitive functions (Smorgon et al. 2004).

3.2 Rodents

Epidemiologic studies conducted on humans do not always show a relationship between neurobehavioral problems and aetiologic factors, because it is difficult to control all of the environmental parameters involved. However, certain genetic parameters can be controlled in experimental animal models. The nature of behavioral alterations can be quantified in properly conceived and conducted neurobehavioral experiments, and such tests are capable of identifying specific toxic effects on the central nervous system. Rodents, in particular, have been used in several types of experiments to study behavioral responses. As a general rule, neurocognitive deficits are more important in juveniles than in adults.

3.2.1 Uranium

The presence of uranium in cerebral structures may have significant consequences for the central nervous system, although most effects derive from chemo- rather than radiotoxicity. The excitability of the hippocampal neurons of rats was decreased 6 months after exposure to depleted uranium. The authors suggest that the hippocampus could participate in cognitive deficits, because it is implicated in learning and memories (Pellmar et al. 1999). Several behavioral studies on rodents showed alterations in cognitive behavior and affected locomotor capacity after chronic or acute exposures to depleted uranium. Alteration of spatial memory and learning was observed in rats after 3 months of exposure by ingestion (Albina et al. 2005; Belles et al. 2005). In another study, hyperactivity in rats was observed after short (2 weeks) and longer (6 months) exposures (Briner and Murray 2005). Sensory-motor deficits in rats were also observed in yet another study (Abou-Donia et al. 2002). After an acute exposure to depleted uranium, rats showed a modification in feeding and rapid eye movement sleep (REM sleep), suggesting respective damage to the hypothalamus and hippocampus had occurred (Lestaevel et al. 2005a).

During chronic exposure to 40 mg/L via drinking water, enriched uranium (EU), in contrast to depleted uranium (DU), provoked a disorder of the sleep–wake cycle in adult male rats. An increase of the REM sleep, after exposures of 1, 1.5 or 2 months, was observed only with EU, but this effect disappeared when the exposure was prolonged to 3 months (Houpert et al. 2005; Lestaevel et al. 2005b). In addition, the spatial working memory and the anxiety of rats were increased by an exposure of 1.5 months to EU, whereas DU had no effect (Houpert et al. 2005). Other experimental results showed that chronic exposure to DU induced effects that counteracted oxidative stress and produced an increase of antioxidant agents in rat brain. In contrast, EU decreased these effects, and EU, but not DU, increased lipid peroxidation (Lestaevel et al. 2009). The authors suggested that the chemical activity of uranium induces a compensatory response to limit oxidative stress, and radiological activity of uranium facilitates, or at least does not inhibit, this oxidative stress. This result could explain the different behavioral results mentioned above that were obtained from DU and EU studies (Houpert et al. 2005; Lestaevel et al. 2005b). All of these results indicate that the rat brain presents differential sensitivity to uranium that depends on the origin of the toxic effect (i.e., chemo- or radiotoxicity).

3.2.2 Caesium

Until recently, few studies have focused on the effect of ^{137}Cs on the central nervous system of mammals. One report showed that the behavior (locomotion, short-term memory) of adult, healthy rats was not disrupted by chronic exposure to ^{137}Cs (6,500 Bq/L) in drinking water (Houpert et al. 2007). However, slight and transitory modifications of the sleep–wake cycle and of electro-encephalographic (EEG) activity were observed in these rats (Lestaevel et al. 2006).

3.2.3 Non-radioactive Cadmium

Experimental studies on rats indicate that early exposure to cadmium can induce behavioral and neurotoxic effects, including a decrease of locomotor activity or an increase of anxiety-like behavior (Baranski 1984; Leret et al. 2003; Oskarsson et al. 1998). Cadmium exposure may also induce a general depression in rats (Ali et al. 1990).

3.3 Fish and Wildlife Species

3.3.1 Behavioral Measurements of Interest

Invertebrates are commonly used for the routine evaluation of the toxicity of chemicals. The behavioral endpoints that are measured include avoidance, feeding depression, valve closure and behavior, among others. If avoidance of contaminants occurs under natural conditions, then bioassays that require forced exposure prevent monitoring of the potential impairments from normal avoidance behavior. Properly conceived avoidance assays may therefore be needed to obtain cost-effective and ecologically relevant information on such behavior.

Chronic feeding assays offer a rapid, cheap and effective tool to garner useful biomonitoring results for contaminants in environmental species (Zhou et al. 2008). Various behavioral patterns of molluscs have been studied in such biomonitoring. In mussels, the periodicity of pumping activity and rest is a sensitive indicator of unfavourable conditions (Salanki et al. 2003). Valve-closure behavior is another useful toxicity endpoint, as has been shown for the Asiatic clam, Corbicula fluminea, exposed to cadmium (Tran et al. 2003). Earthworms are also commonly used in ecotoxicology studies, because of their burrowing habits and importance in soil habitat function. Avoidance and feeding behavior tests with earthworms have already been implemented in uranium mines, and appear to be sensitive endpoints (Andre et al. 2009; Antunes et al. 2008).

Among vertebrates, fish are considered to be interesting and useful models for behavioral studies. Modifications of swimming behavior may alter the capacity of fish to feed, to avoid a predator or to reproduce. Therefore, swimming behavior in fish is a relevant sublethal indicator of toxicity (Little and Finger 1990). The zebrafish, *Danio rerio*, is considered to be a relevant neurobehavioral model because its nervous system is simpler than that of rodents, which allows the analysis of locomotor and memory capacities (Scalzo and Levin 2004). Simple locomotor behavioral effects are important endpoints when fish are studied, because locomotion is relatively easy to quantify with existing computerized imaging systems (Baatrup 2009); such systems allow the evaluation of locomotor activity, school preference or predator avoidance (Gerlai et al. 2000). Fast-start response (*i.e.*, response observed in the first seconds after a stimulus) has also been analysed in pollution-exposed zebrafish (Dlugos and Rabin 2003). Learning tests have recently been developed for use with zebrafish, and these tests are similar to ones used to evaluate learning in rodents and non-human primates (Carvan Iii et al. 2004; Levin and Chen 2004; Williams et al. 2002). Larger fish such as trout, sea bass and mullet are also commonly used as experimental models to study swimming behavior (Beauvais et al. 2000), feeding behavior (Millot et al. 2008), fast-start response (Lefrancois and Domenici 2006), exploration behavior (Millot et al. 2009) and swimming energetic performance (Lefrancois et al. 2007).

Studies have also been performed in contaminated areas to evaluate the effects of toxicants on migratory patterns and preferred nesting sites in birds (Moller and Mousseau 2006).

3.3.2 Uranium

Very few studies have addressed the effects of uranium on aquatic species. Of the studies that have been performed on such species, all were conducted on depleted uranium, which addresses only chemotoxicity, not radiotoxicity. Uranium exposure decreased valve opening time in the molluscs *Corbicula fluminea* (Fournier et al. 2004; Tran et al. 2005) and *Velesunio angasi* (Markich et al. 2000). Burrowing activity of the sludge worm, *Tubifex tubifex*, measured by the length of gallery network in sediment, was also reduced by uranium (Lagauzere et al. 2009). In zebrafish, uranium induced a decrease in fertility and altered courtship behavior (a qualitative measurement) (Bourrachot et al. submitted).

3.3.3 Caesium

In laboratory experiments, avian embryos irradiated with γ rays (2–10 Gy) showed no behavioral changes in post-hatching approach or colour preferences, when compared with controls; this suggests that the nervous system of bird embryos is less susceptible than that of the developing mammalian nervous system to the effects of ionizing radiation (Oppenheim et al. 1970). Similarly, learning and memory abilities of exercised Japanese quails were not modified after single or repeated X-irradiation of the head (Konermann 1970).

Most knowledge on the effects of radioactive caesium on wildlife species comes from field studies performed in the Chernobyl-accident zone. Rodents from the Chernobyl zone, including the vole, *Microtus oeconomus,* showed higher vertical and horizontal locomotion compared to rodents from a less contaminated zone; in addition, the level of the vole's emotional reaction was lower (Karpenko et al. 2003). Radioactive contamination from Chernobyl also had negative effects on many bird species as shown both by their reduced species richness and abundance (Moller and Mousseau 2007b). These effects may result from a preference for nest sites in uncontaminated areas: one study showed that birds, including the great tit, *Parus major,* and the pied flycatcher, *Ficedula hypoleuca,* discriminated against breeding sites that had high radiation dose rates, thereby avoiding radioactively contaminated areas as reproduction sites (Moller and Mousseau 2007a). Furthermore, passerine birds like the barn swallow, *Hirundo rustica,* that breed in contaminated areas, had reduced hatching success and fecundity, and reduced survival prospects as a result of the Chernobyl accident (Moller et al. 2005). The abundance of birds of prey was also reduced in contaminated areas, and there is evidence of a recent increase in the abundance of raptors in less contaminated areas (Moller and Mousseau 2009). Therefore, populations that breed in radioactive sites seem to maintain population levels by immigration from elsewhere, rendering Chernobyl an ecological sink for such immigrants, as was demonstrated by the use of isotope dynamics technique in bird feathers (Moller et al. 2006).

3.3.4 Non-radioactive Cadmium

The effects of cadmium on environmental species are very well documented and several studies have addressed animal behavior endpoints after exposure to this pollutant.

Cadmium altered the feeding behavior of crustaceans after several days of exposure to low contaminant concentrations (6–60 μg/L); the exposed animals were freshwater decapods and amphipods and a terrestrial isopod (Felten et al. 2008; Pestana et al. 2007; Pynnonen 1996).

Molluscs and insects also demonstrated impairment of feeding behavior after cadmium exposure. Feeding rates were decreased from cadmium exposure levels of 0.1–0.5 mg/L in the freshwater snail, *Lymnaea peregra* (Crichton et al. 2004), the land snail, *Helix engaddensis* (Swaileh and Ezzughayyar 2000), and in the midge larvae, *Glyptotendipes pallens* (Heinis et al. 1990). The sandy shore scavenging gastropod, *Nassarius festivus,* showed a decrease in the number of animals feeding after exposure to 0.5 mg/L of cadmium, and the time spent feeding was increased (Cheung et al. 2002). Annelids seem to be more resistant to cadmium: the feeding behavior of three polychaetes was not modified by exposure to 40 mg/kg of cadmium in sediment (Olla et al. 1988). Cadmium also decreased filtration activity of the bivalves *Macoma balthica* (Duquesne et al. 2004) and *Potomida littoralis*

(Mouabad et al. 2001). Cadmium levels exceeding 2.0 mg/L decreased the time of valve opening in the Mediterranean mussel, *Mytilus galloprovincialis* (Ait Fdil et al. 2006).

Cadmium exposure may also alter organism mobility and burrowing activity. The swimming velocity of crustaceans, including the striped barnacle larvae, *Balanus amphitrite* (Lam et al. 2000), the shrimp *Hippolyte inermis* (Untersteiner et al. 2005), the amphipod *Gammarus pulex* (Felten et al. 2008) and the mysid *Neomysis integer* (Roast et al. 2001), was decreased in the presence of cadmium. In benthic animals, the presence of cadmium in sediment impaired the burrowing behavior, as was shown for the bivalves, *Macoma balthica* (McGreer 1979), *Cardium edule* (Amiard and Amiard-Triquet 1986) and the truncated wedgeshell, *Donax trunculus* (Neuberger-Cywiak et al. 2003), and the earthworm, *Lambito mauritü* (Sivakumar et al. 2003). However, cadmium had no effect on burrowing of polychaetes (Olla et al. 1988).

Cadmium potentially induces a displacement of terrestrial animals from their optimum environment. The isopod *Oniscus asellus* avoided cadmium-contaminated food pellets (Zidar et al. 2003). Cadmium also induced a transient avoidance in the whiteworm, *Enchytraeus albidus* (Amorim et al. 2008). Finally, in one study, it was demonstrated that the snail *Physella columbiana* has evolved to detect and avoid heavy metals, including cadmium, at mining sites (Lefcort et al. 2004).

Several authors have also shown the effects of cadmium on fish behavior. In adult zebrafish, high concentrations of cadmium induced a general lethargy (Grillitsch et al. 1999). Adult animals, continuously exposed to cadmium from the embryo stage, presented a decreased escape response to alarm substances, showing that cadmium could alter zebrafish neurogenesis and induce irreversible damage to adult behavior (Kusch et al. 2007). In the rainbow trout, *Oncorhynchus mykiss*, cadmium eliminated antipredatory behavior (Scott et al. 2003), although these animals could successfully escape from a cadmium-contaminated environment (Hansen et al. 1999b). Cadmium created several types of behavioral impairment in guppies: swimming in an imbalanced manner, capsizing, attaching themselves to the surface, difficulty in breathing and gathering around the ventilation filter (Yilmaz et al. 2004). The fast-start response of the sea bass, *Dicentrarchus labrax*, was also altered at 5 μg/L of cadmium exposure (Faucher et al. 2006). Cadmium also decreased exploration activity of the African snakehead, *Parachanna obscura* (Tawari-Fufeyin et al. 2007).

3.3.5 Cobalt Irradiation and Exposure

Sterile insect technique (SIT) is a pest control strategy that involves sterilizing males by exposing them to ionizing radiation, mostly using ^{60}Co. Under these conditions, the mating behavior of the Southern green stink bug, *Nezara viridula*, was not affected by irradiation at 5 Gy, but its fecundity was reduced (Zunié et al. 2002). The mating behavior of the sweet potato weevil, *Cylas formicarius elegantulus*, also was unaffected during the first week of irradiation at 200 Gy, but was altered later (Kumano et al. 2008). In contrast, at an irradiation dose of 15 Gy impaired mating behavior of the European spruce bark beetle, *Ips typographus*, occurred in males, but the burrowing behavior was only slightly modified (Turcani and Vakula 2007). Although the results and the doses are species-dependent, it can be concluded that irradiation does affect insect reproduction.

^{60}Co irradiation also alters reproduction in other species, such as the Kuruma shrimp, *Penaeus japonicus*, in which a decrease in maturation and spawning was shown in females (Sellars et al. 2007).

Fish are able to detect the presence of cobalt in the environment: rainbow trout, *Oncorhynchus mykiss*, and the Chinook salmon, *O. tshawytscha*, escaped from a contaminated area to an uncontaminated zone when cobalt reached levels of 180 and 24 $\mu g/L$, respectively (Hansen et al. 1999a).

4 Consequences

4.1 Consequences in Humans: Neurodegenerative Pathologies

The study of metal and radionuclide effects on neuronal physiology deserves specific attention, because of their consequences on locomotor and cognitive performances and their probable roles in the onset and/or progression of neurodegenerative diseases. Moreover, several neurological diseases (e.g., Alzheimer's and Parkinson's diseases) are characterized by the presence of intra- or extra-cellular deposits that contain proteins associated with metals. The aged people, infants and children are particularly sensitive to these pollutants. Any deficit or excess can alter cell fate or survival and lead to a neurodegenerative insult (Block and Calderon-Garciduenas 2009; Grandjean and Landrigan 2006). Therefore, the understanding of the pathophysiological roles played by metals and radionuclides in brain functions and properties is a public health issue. However, up to the present, the consequences of metals and radionuclides on brain functions and their molecular mechanism of action have been poorly understood.

4.2 Ecological Consequences

Behavioral alterations have biological (e.g., decreases of reproduction or survival) and ecological consequences (e.g., alteration of population structure or ecosystem functioning). Behavior is the link between physiological and ecological processes. Any degradation of the olfactory system in fish could affect migration and food-search behavior, ability to locate or detect spawning beds, and predator avoidance; such effects could lead to animal death and possibly to species extinction (Scott and Sloman 2004).

Ecological consequences may be reversible, if adaptive mechanisms are rapidly manifested. Ecological effects may also be irreversible, when exposure to toxics or irradiation exerts a selection pressure. Such a phenomenon is a basic hypothesis of behavior ecotoxicology that derives from neo-Darwinian theory. Real (1994) suggests that ecological phenomena and community organization are immediate consequences of individual actions and behaviors.

However, the demonstration of a direct link between observed effects at the individual level (in laboratory experiments) and effects on natural communities is rarely done, mainly because of the complexity of natural ecosystems.

5 Conclusion and Future Work

Behavior is defined as the physical manifestation of the integrated physiological responses of an animal to its environment. Behavior can be a potentially excellent biomarker in some species for detecting environmental modification. Several neurochemical mechanisms modulate the adjustment of behavioral responses of organisms to environmental stimuli. The effects of metals on humans and animals (rodents, fish) have been relatively well described. However, data on the effects of radionuclides on humans and animals are missing or poorly represented. From our review, we conclude that the behavior of humans and other animals may be affected in marked ways by exposure to radionuclides. Some behavioral effects result from chemotoxicity of the underlying element, whereas others are caused by its radiotoxicity. The effects of radionuclide pollution on sensorial, locomotor or cognitive performance are variable and depend on several factors, including the element to which exposure takes place and its dose, the duration of the contamination, the species exposed and the type of the cerebral functions altered. We also conclude from our review that despite the fact that behavioral biomarkers can be very useful indicators for environmental damage, there are too few publications available to draw definitive conclusions for most radionuclides and species. Therefore, if the goal is to identify the most probable specific modes of action and damages caused by different radionuclides, further work is needed to improve our knowledge of the brain structures affected by different radionuclides.

6 Summary

Concomitant with the expansion of the nuclear industry, the concentrations of several pollutants, radioactive or otherwise, including uranium, caesium, cadmium and cobalt, have increased over the last few decades. These elemental pollutants do exist in the environment and are a threat to many organisms.

Behavior represents the integration of all the anatomical adaptations and physiological processes that occur within an organism. Compared to other biological endpoints, the effects of pollutants on animal behavior have been the focus of only a few studies. However, behavioral changes appear to be ideal for assessing the effects of pollutants on animal populations, because behavior links physiological functions with ecological processes. The alteration of behavioral responses can have severe implications for survival of individuals and of populations of some species. Behavioral disruptions may derive from several underlying mechanisms: disruption of neuro-sensorial activity and of endocrines, or oxidative and metabolic disruptions. In this review, we presented an overview of the current literature in which the effects of radioactive pollutants on behavior in humans, rodents, fish and wildlife species are addressed. When possible, we have also indicated the potential underlying mechanisms of the behavioral alterations and parameters measured.

In brief, chronic uranium contamination is associated with behavior alterations and mental disorders in humans, and cognitive deficits in rats. Comparative studies on depleted and enriched uranium effects in rats showed that chemical and radiological activities of this metal induced negative effects on several behavioral parameters and also produced brain oxidative stress. Uranium exposure also modifies feeding behavior of bivalves and reproductive behavior of fish.

Studies of the effects of the Chernobyl accident shows that chronic irradiation to ^{137}Cs induces both nervous system diseases and mental disorders in humans leading to increased suicides, as well as modification of preferred nesting sites, reduced hatching success and fecundity in birds that live in the Chernobyl zone. No significant effect from caesium exposure was shown in laboratory experiments with rats, but few studies were conducted.

Data on radioactive cadmium are not available in the literature, but the effects of its metallic form have been well studied. Cadmium induces mental retardation and psychomotor alterations in exposed populations and increases anxiety in rats, leading to depression. Cadmium exposure also results in well-documented effects on feeding and burrowing behavior in several invertebrate species (crustaceans, gastropods, annelids, bivalves) and on different kinds of fish behavior (swimming activity, fast-start response, antipredatory behavior).

Cobalt induces memory deficits in humans and may be involved in Alzheimer's disease; gamma irradiation by cobalt also decreases fecundity and alters mating behavior in insects.

Collectively, data are lacking or are meagre on radionuclide pollutants, and a better knowledge of their actions on the cellular and molecular mechanisms that control animal behavior is needed.

Acknowledgments The authors are very grateful to Prof. Tom Hinton for his reading and improving this manuscript. The work was partly supported by the French National Research Agency, the National Center for Scientific Research (CNRS) and by the ENVIRHOM research program supported by the Institute for Radioprotection and Nuclear Safety (IRSN).

References

Abou-Donia MB, Dechkovskaia AM, Goldstein LB, Shah DU, Bullman SL, Khan WA (2002) Uranyl acetate-induced sensorimotor deficit and increased nitric oxide generation in the central nervous system in rats. Pharmacol Biochem Behav 72(4): 881–90.

Ait Fdil M, Mouabad A, Outzourhit A, Benhra A, Maarouf A, Pihan JC (2006) Valve movement response of the mussel *Mytilus galloprovincialis* to metals (Cu, Hg, Cd and Zn) and phosphate industry effluents from Moroccan Atlantic coast. Ecotoxicology 15(5): 477–486.

Albina ML, Belles M, Linares V, Sanchez DJ, Domingo JL (2005) Restraint stress does not enhance the uranium-induced developmental and behavioral effects in the offspring of uranium-exposed male rats. Toxicology 215(1–2): 69–79.

Ali MM, Mathur N, Chandra SV (1990) Effect of chronic cadmium exposure on locomotor behavior of rats. Indian J Exp Biol 28(7): 653–6.

Ali MM, Shukla GS, Srivastava RS, Mathur N, Chandra SV (1993) Effects of vitamin E on cadmium-induced locomotor dysfunctions in rats. Vet Hum Toxicol 35(2): 109–11.

Amiard JC, Amiard-Triquet C (1986) Influence de différents facteurs écologiques et de contaminations métalliques expérimentales sur le comportement d'enfouissement de *Cardium edule* L. (Mollusques Lamellibranches). Water Air Soil Pollut 27(1–2): 117–130.

Amorim MJB, Novais S, Römbke J, Soares AMVM (2008) Avoidance test with *Enchytraeus albidus* (Enchytraeidae): Effects of different exposure time and soil properties. Environ Pollut 155(1): 112–116.

Andre A, Antunes SC, Gonçalves F, Pereira R (2009) Bait-lamina assay as a tool to assess the effects of metal contamination in the feeding activity of soil invertebrates within a uranium mine area. Environ Pollut 157(8–9): 2368–77.

Antunes SC, Castro BB, Pereira R, Gonçalves F (2008) Contribution for tier 1 of the ecological risk assessment of Cunha Baixa uranium mine (Central Portugal): II. Soil ecotoxicological screening. Sci Tot Environ 390(2–3): 387–395.

Baatrup E (1991) Structural and functional effects of heavy metals on the nervous system, including sense organs, of fish. Comp Biochem Physiol C Pharmacol Toxicol Endocrinol 100(1–2): 253–257.

Baatrup E (2009) Measuring complex behavior patterns in fish – Effects of endocrine disruptors on the guppy reproductive behavior. Hum Ecol Risk Assess 15(1): 53–62.

Baranski B (1984) Behavioral alterations in offspring of female rats repeatedly exposed to cadmium oxide by inhalation. Toxicol Lett 22(1): 53–61.

Barillet S, Adam C, Palluel O, Devaux A (2007) Bioaccumulation, oxidative stress, and neurotoxicity in *Danio rerio* exposed to different isotopic compositions of uranium. Environ Toxicol Chem 26(3): 497–505.

Beauvais SL, Jones SB, Brewer SK, Little EE (2000) Physiological measures of neurotoxicity of diazinon and malathion to larval rainbow trout (*Oncorhynchus mykiss*) and their correlation with behavioral measures. Environ Toxicol Chem 19(7): 1875–1880.

Beauvais SL, Jones SB, Parris JT, Brewer SK, Little EE (2001) Cholinergic and behavioral neurotoxicity of carbaryl and cadmium to larval rainbow trout (*Oncorhynchus mykiss*). Ecotoxicol Environ Saf 49(1): 84–90.

Belles M, Albina ML, Linares V, Gomez M, Sanchez DJ, Domingo JL (2005) Combined action of uranium and stress in the rat. I. Behavioral effects. Toxicol Lett 158(3): 176–185.

Bensoussan H, Grancolas L, Dhieux-Lestaevel B, Delissen O, Vacher CM, Dublineau I, Voisin P, Gourmelon P, Taouis M, Lestaevel P (2009) Heavy metal uranium affects the brain cholinergic system in rat following sub-chronic and chronic exposure. Toxicology 261(1–2): 59–67.

Beyers DW, Farmer MS (2001) Effects of copper on olfaction of Colorado pikeminnow. Environ Toxicol Chem 20(4): 907–912.

Blechinger SR, Kusch RC, Haugo K, Matz C, Chivers DP, Krone PH (2007) Brief embryonic cadmium exposure induces a stress response and cell death in the developing olfactory system followed by long-term olfactory deficits in juvenile zebrafish. Toxicol Appl Pharmacol 224(1): 72–80.

Block ML, Calderon-Garciduenas L (2009) Air pollution: Mechanisms of neuroinflammation and CNS disease. Trends Neurosci 32(9): 506–516.

Bone Q, Moore RH (2008) Behavior and Cognition. In: Francis T (ed) Biology of fishes, 3rd ed. Taylor & Francis Group, Abingdon, pp 409–436.

Bourrachot S, Brion F, Palluel O, Adam-Guillermin C, Giblin R (submitted) Effect of uranium on zebrafish reproduction: Measurement of genotoxicity and vitellogenin. Aquat Toxicol.

Brewer SK, Little EE, DeLonay AJ, Beauvais SL, Jones SB, Ellersieck MR (2001) Behavioral dysfunctions correlate to altered physiology in rainbow trout (*Oncorynchus mykiss*) exposed to cholinesterase-inhibiting chemicals. Arch Environ Contam Toxicol 40(1): 70–76.

Briner W, Murray J (2005) Effects of short-term and long-term depleted uranium exposure on open-field behavior and brain lipid oxidation in rats. Neurotoxicol Teratol 27(1): 135–144.

Bussy C, Lestaevel P, Dhieux B, Amourette C, Paquet F, Gourmelon P, Houpert P (2006) Chronic ingestion of uranyl nitrate perturbs acetylcholinesterase activity and monoamine metabolism in male rat brain. Neurotoxicology 27(2): 245–252.

Carvan Iii MJ, Loucks E, Weber DN, Williams FE (2004) Ethanol effects on the developing zebrafish: Neurobehavior and skeletal morphogenesis. Neurotoxicol Teratol 26(6): 757–768.

Cheung SG, Tai KK, Leung CK, Siu YM (2002) Effects of heavy metals on the survival and feeding behavior of the sandy shore scavenging gastropod *Nassarius festivus* (Powys). Mar Pollut Bull 45(1–12): 107–113.

Chivers DP, Smith RJF (1998) Chemical alarm signalling in aquatic predator-prey systems: A review and prospectus. Ecoscience 5(3): 338–352.

Clotfelter ED, Bell AM, Levering KR (2004) The role of animal behavior in the study of endocrine-disrupting chemicals. Anim Behav 68(4): 665–676.

Collin SP (2007) Nervous and sensory systems. In: McKenzie DJ, Farrell AP, Brauner CJ (eds) Primitive fishes. Fish physiology series, vol. 26. Academic Press, Amsterdam, pp 121–179.

Comeau Y, Brisson J, Reville JP, Forget C, Drizo A (2001) Phosphorus removal from trout farm effluents by constructed wetlands. Water Sci Technol 44(11–12): 55–60.

Cousin X, Strahle U, Chatonnet A (2005) Are there non-catalytic functions of acetyl-cholinesterases? Lessons from mutant animal models. Bioessays 27(2): 189–200.

Crichton CA, Conrad AU, Baird DJ (2004) Assessing stream grazer response to stress: a post-exposure feeding bioassay using the freshwater snail *Lymnaea peregra* (Muller). Bull Environ Contam Toxicol 72(3): 564–570.

Dlugos CA, Rabin RA (2003) Ethanol effects on three strains of zebrafish: Model system for genetic investigations. Pharmacol Biochem Behav 74(2): 471–480.

Døving KB, Lastein S (2009) The alarm reaction in fishes: odorants, modulations of responses, neural pathways. Ann NY Acad Sci (International Symposium on Olfaction and Taste) 1170: 413–423.

Duquesne S, Liess M, Bird DJ (2004) Sub-lethal effects of metal exposure: Physiological and behavioral responses of the estuarine bivalve *Macoma balthica*. Mar Environ Res 58(2–5): 245–250.

Faucher K, Fichet D, Miramand P, Lagardere JP (2006) Impact of acute cadmium exposure on the trunk lateral line neuromasts and consequences on the "C-start" response behavior of the sea bass (*Dicentrarchus labrax* L.; Teleostei, Moronidae). Aquat Toxicol 76(3–4): 278–294.

Felten V, Charmantier G, Mons R, Geffard A, Rousselle P, Coquery M, Garric J, Geffard O (2008) Physiological and behavioral responses of *Gammarus pulex* (Crustacea: Amphipoda) exposed to cadmium. Aquat Toxicol 86(3): 413–425.

Fournier E, Tran D, Denison F, Massabuau JC, Garnier-Laplace J (2004) Valve closure response to uranium exposure for a freshwater bivalve (*Corbicula fluminea*): Quantification of the influence of pH. Environ Toxicol Chem 23(5): 1108–1114.

Gerlai R, Lahav M, Guo S, Rosenthal A (2000) Drinks like a fish: zebra fish (*Danio rerio*) as a behavior genetic model to study alcohol effects. Pharmacol Biochem Behav 67(4): 773–782.

Goasguen J, Lapresle J, Ribot C, Rocquet G (1982) Chronic neurological syndrome resulting from intoxication with metallic uranium. Nouv Presse Med 11(2): 119–121.

Grandjean P, Landrigan PJ (2006) Developmental neurotoxicity of industrial chemicals. Lancet 368(9553): 2167–2178.

Grillitsch B, Vogl C, Wytek R (1999) Qualification of spontaneous undirected locomotor behavior of fish for sublethal toxicity testing. Part II. Variability of measurement parameters under toxicant-induced stress. Environ Toxicol Chem 18(12): 2743–2750.

Grue CE, Gardner SC, Gibert PL (2002) On the significance of pollutant-induced alterations in the behavior of fish and wildlife. In: Dell'Omo G (ed) Behavioral ecotoxicology. Ecotoxicology and environmental toxicology series. Wiley, Chichester, pp 1–90.

Hansen JA, Marr JCA, Lipton J, Cacela D, Bergman HL (1999a) Differences in neurobehavioral responses of chinook salmon (*Oncorhynchus tshawytscha*) and rainbow trout (*Oncorhynchus mykiss*) exposed to copper and cobalt: Behavioral avoidance. Environ Toxicol Chem 18(9): 1972–1978.

Hansen JA, Woodward DF, Little EE, Delonay AJ, Bergman HL (1999b) Behavioral avoidance: Possible mechanism for explaining abundance and distribution of trout species in a metal-impacted river. Environ Toxicol Chem 18(2): 313–317.

Hart RP, Rose CS, Hamer RM (1989) Neuropsychological effects of occupational exposure to cadmium. J Clin Exp Neuropsychol 11(6): 933–943.

Heinis F, Timmermans KR, Swain WR (1990) Short-term sublethal effects of cadmium on the filter feeding chironomid larva *Glyptotendipes pallens* (Meigen) (Diptera). Aquat Toxicol 16(1): 73–86.

Hertz Y, Madar Z, Hepher B, Gertler A (1989) Glucose metabolism in the common carp (*Cyprinus carpio* L.): the effects of cobalt and chromium. Aquaculture 76(3–4): 255–267.

Houpert P, Lestaevel P, Bussy C, Paquet F, Gourmelon P (2005) Enriched but not depleted uranium affects central nervous system in long-term exposed rat. Neurotoxicology 26(6): 1015–1020.

Houpert P, Bizot JC, Bussy C, Dhieux B, Lestaevel P, Gourmelon P, Paquet F (2007) Comparison of the effects of enriched uranium and 137-cesium on the behavior of rats after chronic exposure. Int J Radiat Biol 83(2): 99–104.

Howland JW (1949) Studies on human exposures to uranium compounds. In: Voegtlin C, Hodge HC (eds) Pharmacology and toxicology of uranium. McGraw-Hill Book Company, New York, NY, pp 993–1017.

IUPAC (2004) International Union for Pure and Applied Chemistry (IUPAC) stability constant database. http://www.acadsoft.co.uk.

Ivanov VK, Maksioutov MA, Chekin S, Kruglova ZG, Petrov AV, Tsyb AF (2000) Radiation-epidemiological analysis of incidence of non-cancer diseases among the Chernobyl liquidators. Health Phys 78(5): 495–501.

Jiang HM, Han GA, He ZL (1990) Clinical significance of hair cadmium content in the diagnosis of mental retardation of children. Chin Med J 103(4): 331–334.

Jordan CM, Whitman RD, Harbut M (1997) Memory deficits and industrial toxicant exposure: a comparative study of hard metal, solvent and asbestos workers. Int J Neurosci 90(1–2): 113–128.

Kamarli Z, Abdulina A (1996) Health conditions among workers who participated in the cleanup of the Chernobyl accident. World Health Stat Q 49(1): 29–31.

Karpenko NA, Buntova EG, Alesina NY, Lyabik VV (2003) Estimation of long-time effects of Chernobyl NPP accident on behavior markers in a little rodent populations. Radiatsionnaya Biologiya. Radioekologiya 43(6): 682–687.

Kavlock RJ, Daston GP, DeRosa C, Fenner-Crisp P, Gray LE, Kaattari S, Lucier G, Luster M, Mac MJ, Maczka C, Miller R, Moore J, Rolland R, Scott G, Sheehan DM, Sinks T, Tilson HA (1996) Research needs for the risk assessment of health and environmental effects of endocrine disruptors: a report of the U.S. EPA-sponsored workshop. Environ Health Perspect 104(Suppl 4): 715–740.

Konermann G (1970) Learning and memory abilities of exercised Japanese quails (*Coturnix c. japonica*) after irradiation of the head. Strahlentherapie 140(6): 757–764.

Kumano N, Haraguchi D, Kohama T (2008) Effect of irradiation on mating performance and mating ability in the West Indian sweetpotato weevil, *Euscepes postfasciatus*. Entomol Exp Appl 127(3): 229–236.

Kusch R, Krone P, Chivers D (2007) Chronic exposure to low concentrations of water-borne cadmium during embryonic and larval development results in the long-term hindrance of anti-predator behavior in zebrafish. Environ Toxicol Chem 27(3): 705–710.

Lagauzere S, Terrail R, Bonzom JM (2009) Ecotoxicity of uranium to Tubifex tubifex worms (Annelida, Clitellata, Tubificidae) exposed to contaminated sediment. Ecotoxicol Environ Saf 72(2): 527–537.

Lam PKS, Wo KT, Wu RSS (2000) Effects of cadmium on the development and swimming behavior of barnacle larvae *Balanus amphitrite* Darwin. Environ Toxicol 15(1): 8–13.

Lefcort H, Abbott DP, Cleary DA, Howell E, Keller NC, Smith MM (2004) Aquatic snails from mining sites have evolved to detect and avoid heavy metals. Arch Environ Contam Toxicol 46(4): 478–484.

Lefrancois C, Domenici P (2006) Locomotor kinematics and behavior in the escape response of European sea bass, *Dicentrarchus labrax* L., exposed to hypoxia. Mar Biol 149(4): 969–977.

Lefrancois C, Nieto Amat J, Kostecki C, Ferrari R, Domenici P (2007) The effect of oxygen and temperature on the energetics of swimming in *Mugil cephalus*. Comp Biochem Physiol A Mol Integr Physiol 146(4, Supplement 1): S85.

Lerebours A, Bourdineaud J-P, Van der Ven K, Vandenbrouck T, Gonzalez P, Camilleri V, Floriani M, Garnier-Laplace J, Adam-Guillermin C (2010) Sub-lethal effects of waterborne uranium exposures on the zebrafish brain: transcriptional responses and alterations of the olfactory bulb ultrastructure. Environ Sci Technol 44: 1438–1443.

Leret ML, Millan JA, Antonio MT (2003) Perinatal exposure to lead and cadmium affects anxiety-like behavior. Toxicology 186(1–2): 125–130.

Lestaevel P, Houpert P, Bussy C, Dhieux B, Gourmelon P, Paquet F (2005a) The brain is a target organ after acute exposure to depleted uranium. Toxicology 212(2–3): 219–226.

Lestaevel P, Bussy C, Paquet F, Dhieux B, Clarencon D, Houpert P, Gourmelon P (2005b) Changes in sleep-wake cycle after chronic exposure to uranium in rats. Neurotoxicol Teratol 27(6): 835–840.

Lestaevel P, Dhieux B, Tourlonias E, Houpert P, Paquet F, Voisin P, Aigueperse J, Gourmelon P (2006) Evaluation of the effect of chronic exposure to 137Cesium on sleep-wake cycle in rats. Toxicology 226(2–3): 118–125.

Lestaevel P, Romero E, Dhieux B, Ben Soussan H, Berradi H, Dublineau I, Voisin P, Gourmelon P (2009) Different pattern of brain pro-/anti-oxidant activity between depleted and enriched uranium in chronically exposed rats. Toxicology 258(1): 1–9.

Levin ED, Chen E (2004) Nicotinic involvement in memory function in zebrafish. Neurotoxicol Teratol 26(6): 731–735.

Linney E, Upchurch L, Donerly S (2004) Zebrafish as a neurotoxicological model. Neurotoxicol Teratol 26(6): 709–718.

Little EE, Finger SE (1990) Swimming behavior as an indicator of sublethal toxicity in fish. Environ Toxicol Chem 9(1): 13–19.

Markich SJ, Brown PL, Jeffree RA, Lim RP (2000) Valve movement responses of *Velesunio angasi* (Bivalvia: Hyriidae) to manganese and uranium: an exception to the free ion activity model. Aquat Toxicol 51(2): 155–175.

Matz CJ, Krone PH (2007) Cell death, stress-responsive transgene activation, and deficits in the olfactory system of larval zebrafish following cadmium exposure. Environ Sci Technol 41(14): 5143–5148.

McDiarmid MA, Keogh JP, Hooper FJ, McPhaul K, Squibb K, Kane R, DiPino R, Kabat M, Kaup B, Anderson L, Hoover D, Brown L, Hamilton M, Jacobson-Kram D, Burrows B, Walsh M (2000) Health effects of depleted uranium on exposed Gulf War veterans. Environ Res 82(2): 168–180.

McGreer ER (1979) Sublethal effects of heavy metal contaminated sediments on the bivalve *Macoma balthica* (L.). Mar Pollut Bull 10(9): 259–262.

McGeer JC, Szebedinszky C, McDonald DG, Wood CM (2000) Effects of chronic sublethal exposure to waterborne Cu, Cd or Zn in rainbow trout. 1: Iono-regulatory disturbance and metabolic costs. Aquat Toxicol 50(3): 231–243.

Migliore L, Coppedè F (2009) Environmental-induced oxidative stress in neurodegenerative disorders and aging. Mutat Res – Genet Toxicol Environ Mutagen 674(1–2): 73–84.

Millot S, Bégout ML, Chatain B (2009) Exploration behavior and flight response toward a stimulus in three sea bass strains (*Dicentrarchus labrax* L.). Appl Anim Behav Sci 119(1–2): 108–114.

Millot S, Bégout ML, Person-Le Ruyet J, Breuil G, Di-Poï C, Fievet J, Pineau P, Roué M, Sévère A (2008) Feed demand behavior in sea bass juveniles: Effects on individual specific growth rate variation and health (inter-individual and inter-group variation). Aquaculture 274(1): 87–95.

Moller AP, Mousseau TA (2006) Biological consequences of Chernobyl: 20 years on. Trends Ecol Evol 21(4): 200–207.

Moller AP, Mousseau TA (2007a) Birds prefer to breed in sites with low radioactivity in Chernobyl. Proc R Soc B Biol Sci 274(1616): 1443–1448.

Moller AP, Mousseau TA (2007b) Species richness and abundance of forest birds in relation to radiation at Chernobyl. Biol Lett 3(5): 483–486.

Moller AP, Mousseau TA (2009) Reduced abundance of raptors in radioactively contaminated areas near Chernobyl. J Ornithol 150(1): 239–246.

Moller AP, Mousseau TA, Milinevsky G, Peklo A, Pysanets E, Szep T (2005) Condition, reproduction and survival of barn swallows from Chernobyl. J Anim Ecol 74(6): 1102–1111.

Moller AP, Hobson KA, Mousseau TA, Peklo AM (2006) Chernobyl as a population sink for barn swallows: Tracking dispersal using stable-isotope profiles. Ecol Appl 16(5): 1696–1705.

Monleau M, Bussy C, Lestaevel P, Houpert P, Paquet F, Chazel V (2005) Bioaccumulation and behavioral effects of depleted uranium in rats exposed to repeated inhalations. Neurosci Lett 390(1): 31–36.

Mouabad A, Ait Fdil M, Maarouf A, Pihan JC (2001) Pumping behavior and filtration rate of the freshwater mussel *Potomida littoralis* as a tool for rapid detection of water contamination. Aquat Ecol 35(1): 51–60.

Nath K, Kumar N (1988) Cobalt induced alterations in the carbohydrate metabolism of a freshwater tropical perch, *Colisa fasciatus*. Chemosphere 17(2): 465–474.

Neuberger-Cywiak L, Achituv Y, Garcia EM (2003) Effects of zinc and cadmium on the burrowing behavior, LC50, and LT50 on *Donax trunculus* linnaeus (Bivalvia-Donacidae). Bull Environ Contam Toxicol 70(4): 713–722.

Nordberg GF, Fowler BA, Nordberg M, Friberg LT (2007) Handbook on the toxicology of metals, 3rd ed. Academic Press/Elsevier, Amsterdam, 996p.

Olla BL, Estelte VB, Swartz RC, Braun G, Studholme AL (1988) Responses of polychaetes to cadmium-contaminated sediment: Comparison of uptake and behavior. Environ Toxicol Chem 7(7): 587–592.

Oppenheim RW, Jones JR, Gottlieb G (1970) Embryonic motility and posthatching perception in birds after prenatal gamma irradiation. J Comp Physiol Psychol 71(1): 6–21.

Oskarsson A, Palminger Hallen I, Sundberg J, Petersson Grawe K (1998) Risk assessment in relation to neonatal metal exposure. Analyst 123(1): 19–23.

Peakall DB (1996) Disrupted patterns of behavior in natural populations as an index of ecotoxicity. Environ Health Perspect 104(suppl 2): 331–335.

Pellmar TC, Keyser DO, Emery C, Hogan JB (1999) Electrophysiological changes in hippocampal slices isolated from rats embedded with depleted uranium fragments. Neurotoxicology 20(5): 785–792.

Pestana JLT, Ré A, Nogueira AJA, Soares AMVM (2007) Effects of cadmium and zinc on the feeding behavior of two freshwater crustaceans: *Atyaephyra desmarestii* (Decapoda) and *Echinogammarus meridionalis* (Amphipoda). Chemosphere 68(8): 1556–1562.

Pynnonen K (1996) Heavy metal-induced changes in the feeding and burrowing behavior of a Baltic isopod, *Saduria* (*Mesidotea*) *entomon* L. Mar Environ Res 41(2): 145–156.

Real LA (1994) Behavioral mechanisms in evolutionary ecology. University of Chicago Press, Chicago, USA. 469p.

Richards JG, Curtis PJ, Burnison BK, Playle RC (2001) Effects of natural organic matter source on reducing metal toxicity to rainbow trout (*Oncorhynchus mykiss*) and on metal binding to their gills. Environ Toxicol Chem 20(6): 1159–1166.

Roast SD, Widdows J, Jones MB (2001) Impairment of mysid (*Neomysis integer*) swimming ability: an environmentally realistic assessment of the impact of cadmium exposure. Aquat Toxicol 52(3–4): 217–227.

Roex EW, Keijzers R, van Gestel CA (2003) Acetylcholinesterase inhibition and increased food consumption rate in the zebrafish, *Danio rerio*, after chronic exposure to parathion. Aquat Toxicol 64(4): 451–60.

Salanki J, Hiripi L (1990) Effect of heavy metals on the serotonin and dopamine systems in the central nervous system of the freshwater mussel (*Anodonta cygnea* L.). Comp Biochem Physiol C Pharmacol Toxicol Endocrinol 95(2): 301–305.

Salanki J, Budai D, Hiripi L, Kasa P (1993) Acetylcholine level in the brain and other organs of the bivalve *Anodonta cygnea* L. and its modification by heavy metals. Acta Biol Hung 44(1): 21–24.

Salanki J, Farkas A, Kamardina T, Rozsa KS (2003) Molluscs in biological monitoring of water quality. Toxicol Lett 140–141: 403–410.

Scalzo FM, Levin ED (2004) The use of zebrafish (*Danio rerio*) as a model system in neurobehavioral toxicology. Neurotoxicol Teratol 26(6): 707–718.

Scott GR, Sloman KA (2004) The effects of environmental pollutants on complex fish behavior: integrating behavioral and physiological indicators of toxicity. Aquat Toxicol 68(4): 369–392.

Scott GR, Sloman KA, Rouleau C, Wood CM (2003) Cadmium disrupts behavioral and physiological responses to alarm substance in juvenile rainbow trout (*Oncorhynchus mykiss*). J Exp Biol 206(Pt 11): 1779–1790.

Sellars MJ, Coman GJ, Callaghan TR, Arnold SJ, Wakeling J, Degnan BM, Preston NP (2007) The effect of ionizing irradiation of post-larvae on subsequent survival and reproductive performance in the Kuruma shrimp, *Penaeus (Marsupenaeus) japonicus* (Bate). Aquaculture 264(1–4): 309–322.

Sivakumar S, Kavitha K, Sivaraj R, Prabha D, Subburam V (2003) Effect of cadmium and mercury on the survival morphology and burrowing behavior of the earthworm *Lambito mauritii* (Kinberg). Ind J Environ Protect 23(7): 792–799.

Smorgon C, Mari E, Atti AR, Dalla Nora E, Zamboni PF, Calzoni F, Passaro A, Fellin R (2004) Trace elements and cognitive impairment: an elderly cohort study. Arch Gerontol Geriatr (S9): 393–402.

Sulkowski WJ, Rydzewski B, Miarzynska M (2000) Smell impairment in workers occupationally exposed to cadmium. Acta Oto-Laryngol 120(2): 316–318.

Swaileh KM, Ezzughayyar A (2000) Effects of dietary Cd and Cu on feeding and growth rates of the landsnail *Helix engaddensis*. Ecotoxicol Environ Saf 47(3): 253–260.

Tallkvist J, Persson E, Henriksson J, Tjalve H (2002) Cadmium-metallothionein interactions in the olfactory pathways of rats and pikes. Toxicol Sci 67(1): 108–113.

Tawari-Fufeyin P, Opute P, Ilechie I (2007) Toxicity of cadmium to *Parachanna obscura*: As evidenced by alterations in hematology, histology, and behavior. Toxicol Environ Chem 89(2): 243–248.

Tomasek L, Swerdlow AJ, Darby SC, Placek V, Kunz E (1994) Mortality in uranium miners in west Bohemia: a long-term cohort study. Occup Environ Med 51(5): 308–315.

Tournier BB, Frelon S, Tourlonias E, Agez L, Delissen O, Dublineau I, Paquet F, Petitot F (2009) Role of the olfactory receptor neurons in the direct transport of inhaled uranium to the rat brain. Toxicol Lett 190(1): 66–73.

Tran D, Ciret P, Ciutat A, Durrieu G, Massabuau JC (2003) Estimation of potential and limits of bivalve closure response to detect contaminants: application to cadmium. Environ Toxicol Chem 22(4): 914–920.

Tran D, Bourdineaud JP, Massabuau JC, Garnier-Laplace J (2005) Modulation of uranium bioaccumulation by hypoxia in the freshwater clam *Corbicula fluminea*: Induction of multixenobiotic resistance protein and heat shock protein 60 in gill tissues. Environ Toxicol Chem 24(9): 2278–2284.

Turcani M, Vakula J (2007) The influence of irradiation on the behavior and reproduction success of eight toothed bark beetle *Ips typographus* (Coleoptera: Scolytidae). J For Sci 53(speciss): 31–37.

Untersteiner H, Gretschel G, Puchner T, Napetschnig S, Kaiser H (2005) Monitoring behavioral responses to the heavy metal cadmium in the marine shrimp *Hippolyte inermis* Leach (Crustacea: Decapoda) with video imaging. Zool Stud 44(1): 71–80.

Vetillard A, Bailhache T (2005) Cadmium: an endocrine disrupter that affects gene expression in the liver and brain of juvenile Rainbow trout. Biol Reprod 72(1): 119–126.

Williams FE, White D, Messer WS (2002) A simple spatial alternation task for assessing memory function in zebrafish. Behav Processes 58(3): 125–132.

Yilmaz M, Gül A, Karaköse E (2004) Investigation of acute toxicity and the effect of cadmium chloride (CdCl2 · H2O) metal salt on behavior of the guppy (*Poecilia reticulata*). Chemosphere 56(4): 375–380.

Zala SM, Penn DJ (2004) Abnormal behaviors induced by chemical pollution: A review of the evidence and new challenges. Anim Behav 68(4): 649–664.

Zalups RK, Koropatnick J (2000) Molecular biology and toxicology of metals. Taylor & Francis, London, 603p.

Zhou Q, Zhang J, Fu J, Shi J, Jiang G (2008) Biomonitoring: An appealing tool for assessment of metal pollution in the aquatic ecosystem. Anal Chim Acta 606(2): 135–150.

Zidar P, Kaschl UI, Drobne D, Bozic J, Strus J (2003) Behavioral response in paired food choice experiments with *Oniscus asellus* (Crustacea, Isopoda) as an indicator of different food quality. Arh Hig Rada Toksikol 54(3): 177–181.

Zuniè A, Eokl A, Sersa G (2002) Effects of 5-Gy irradiation on fertility and mating behavior of *Nezara viridula* (Heteroptera: Pentatomidae). Radiol Oncol 36(3) 231–237.

Illicit Drugs: Contaminants in the Environment and Utility in Forensic Epidemiology

Christian G. Daughton

Contents

1 Introduction 60
2 What Is an "Illicit" Drug? 67
  2.1 Terminology 67
  2.2 Differences Between Illicit and Licit Drugs as Environmental Contaminants . . 71
3 The Core Illicit Drugs and the Environment 73
  3.1 Environmental Occurrence 76
  3.2 Adulterants and Impurities as Potential Environmental Contaminants 81
4 Large-Scale Exposure or Source Assessments via Dose Reconstruction 82
  4.1 Sewage Epidemiology or Forensics – FEUDS 83
  4.2 FEUDS for Community-Wide Dose Reconstruction of Illicit Drugs 83
  4.3 Quality Assurance and FEUDS 86
  4.4 Summary of Published Research in FEUDS 86
  4.5 Legal Concerns Surrounding FEUDS 89
5 Illicit Drugs in the Money Supply 90
6 Illicit Drugs in Ambient Air 91
7 Other Routes of Illicit Drug Impact on the Environment 91
  7.1 Clan Labs 91
  7.2 Livestock and Racing Animals 92
  7.3 Dermal Contact and Transfer 93
  7.4 Diversion 93
  7.5 Disposal of Leftover Medications 94
8 Illicit Drugs and Environmental Impact 94
  8.1 Fate and Transport 94
  8.2 Ecotoxicology 95

C.G. Daughton (✉)
Environmental Chemistry Branch, National Exposure Research Laboratory, U.S. Environmental Protection Agency, Las Vegas, NV 89119, USA
e-mail: daughton.christian@epa.gov

D.M. Whitacre (ed.), *Reviews of Environmental Contamination and Toxicology*,
Reviews of Environmental Contamination and Toxicology 210,
DOI 10.1007/978-1-4419-7615-4_3, © Springer Science+Business Media, LLC 2011

9 The Future 96
    9.1 Advancing the Utility of FEUDS 96
    9.2 Real-Time Monitoring of Community-Wide Health and Disease: Using
        Sewage Information Mining (SIM) 98
10 Summary 99
References 101

1 Introduction

The spectrum of chemicals recognized as contributing to widespread contamination of the environment was extended to pharmaceutical ingredients as early as the 1970s. The topic, however, did not begin to attract broader scientific attention until the mid-1990s (Daughton 2009a). Occurring generally at levels below 1 μg/L (1 part per billion) in ambient waters, recognition of the near-ubiquitous presence of pharmaceuticals in a wide variety of environmental compartments serves as a stunning measure of the advancements in analytical chemistry and of our still-emerging understanding of the scope and complexity of xenobiotic occurrence in the environment.

More so than with any other class of environmental contaminants, drugs have served to illustrate the intimate, inseparable, and immediate connections between the actions, activities, and behaviors of individual citizens and the environment in which they live (Daughton 2001a). Drug contaminants also highlight the profound changes that have occurred in how risk is perceived by the public. After all, it has now been 40 years since the occurrence of an emblematic event that was a major catalyst for the creation of the US EPA (in 1971) and which was followed soon after by the Federal Water Pollution Control Amendments of 1972 and the Clean Water Act of 1977 and later by the Water Quality Act of 1987. This event was the 1969 Ohio Cuyahoga River fire, which otherwise had little broad environmental significance because more than a dozen similar fires had occurred in the preceding 100 years (with the largest occurring in 1952), all resulting from the river's continual accumulation of combustible floating debris and petroleum wastes.

Gross-level pollution of waterways had not been confined to the Cuyahoga River. But, the 1969 fire was a landmark event and changed the way the environment was viewed. The extent of progress and effectiveness of pollution regulation, mitigation, control, and prevention over the last 40 years is now reflected by a focus on trace-level chemical pollutants – an evolutionary change not contemplated in the early 1970s but made possible by continual advancements in instrumental analytical chemistry that began in the 1960s. This focus is embodied particularly with the so-called emerging contaminants (Daughton 2009b) and the myriad others not yet noticed or identified, which could be referred to as the "quiet contaminants."

Until the mid-2000s, the emerging study of pharmaceuticals in the environment (PiE) inexplicably excluded from consideration one major aspect – the contributions to overall environmental loadings by the so-called illicit drugs. A structurally diverse group of chemical agents uniformly possessing extremely high potential for biological effects in humans and non-target organisms alike, illicit drugs are used in enormous quantities worldwide. However, the actual magnitude of illicit drugs is unknown and can only be roughly estimated. The potential for illicit drugs to enter the environment via a wide array of pathways should not differ much from that of pharmaceuticals used in the practice of medicine. Although it had been known for many decades that illicit drugs and their metabolites (just as with pharmaceuticals used in the practice of medicine) are excreted in urine, feces, hair, and sweat, the ramifications for the environment were basically ignored until 1999 (Daughton and Ternes 1999) and 2001 (Daughton 2001a, c), when the scope of concerns surrounding PiE was expanded to include illicit drugs. In characterizing and assessing risks incurred from PiE, both licit and illicit drugs need to be considered seamlessly.

Perhaps the first published indication that illicit drugs might be pervasive contaminants of our immediate surroundings and the environment was a 1987 FBI study performed in response to a newspaper report 2 years earlier that cocaine was present on money in general circulation (Aaron and Lewis 1987). Over the intervening 20 years, analogous surveys of illicit drug ambient contaminants have been attempted for the first time for sewage wastewaters (Khan 2002), surface waters (Zuccato et al. 2005), air (Cecinato and Balducci 2007), sewage sludge (Kaleta et al. 2006), biosolids (Jones-Lepp and Stevens 2007), and most recently drinking water (Huerta-Fontela et al. 2008a). An examination of the US EPA's bibliographic database on pharmaceuticals in the environment (USEPA 2009b) shows that the core journal references having a major focus on illicit drugs in wastewaters, ambient waters, drinking water, or the air total about 70 (excluding those published on the topic of drugs on money). The number of references (in any type of technical publication) dealing with illicit drugs in the environment is fewer than 200; this number comprises only 2% of the roughly 10,000 documents that address the general topic of PiE.

Presented herein is the first broad overview of the topic of illicit drugs as environmental contaminants. Summary perspectives are provided of the published data on their occurrence in a spectrum of environmental compartments, what their occurrence might mean with regard to risk, and an historic perspective on how their occurrence can be used as an analytical measurement tool to assess society-wide usage of illicit drugs. An illustrated flowchart depicting the varied routes by which illicit drugs gain entry to our immediate surroundings and to the ambient environment is presented in Fig. 1.

The chronology of seminal publications that address the significant aspects of illicit drugs and the environment is presented in Table 1. The topic is transdisciplinary, involving the knowledge from a variety of disparate but intersecting fields, including health care, pharmacology, criminology, forensic sciences, epidemiology, toxicology, environmental and analytical chemistry, and sanitary engineering.

This chapter builds upon previous work, which is scheduled to be published in one of the only books to date devoted to the topic of illicit drugs in the environment (Daughton 2011 – in press).

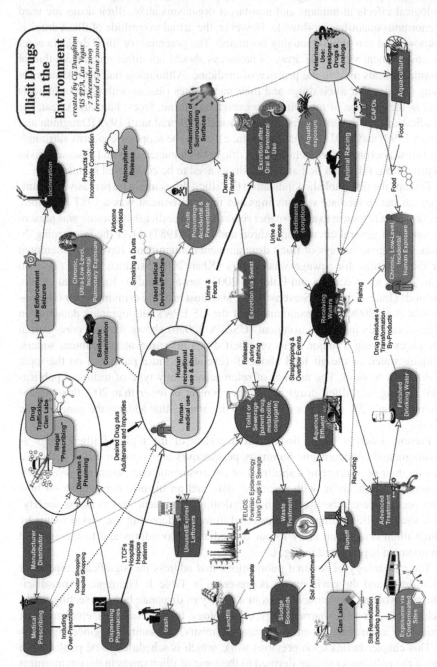

Fig. 1 Illicit drugs in the environment (relative significance of routes indicated by intensity of lines)

Table 1 Chronology of some selected seminal publications regarding illicit drugs in the environment

Year	Aspect	Unique features of study	References
1987	M	*First report in a journal confirming the presence of an illicit drug (cocaine) on banknotes in general circulation) (objective to distinguish "drug" money from "innocent" money)*	Aaron and Lewis (1987)
1998	A	*Perhaps first data on an illicit drug in the ambient environment; non-target analysis revealed cocaine associated with fractions of particulate matter in outdoor air (Los Angeles)*	Hannigan et al. (1998)
2000	M	*First comprehensive overview of drugs on banknotes*	Sleeman et al. (2000)
2001	F	*Use of residues in sewage to reconstruct community-wide drug usage first proposed (later to be termed "sewage epidemiology" or "sewage forensics," or sometimes "community drug testing" or "community urinalysis"); first discussion to broaden the topic of drugs as environmental contaminants to include illicit drugs*	Daughton (2001c)
2002	WW	Morphine, methamphetamine, and methadone in sewage	Khan (2002)
2004	WW, monit	Methamphetamine and MDMA (3,4-methylenedioxymethamphetamine) in WWTP (Wastewater Treatment Plant) effluent; first report by US EPA of illicit drug in the environment; *first use of integrative time-weighted sampling for illicit drugs in wastewaters*	Jones-Lepp et al. (2004)
2004	M	THC (Δ9-tetrahydrocannabinol), cannabinol, and cannabidiol on banknotes from the USA and other countries	Lavins et al. (2004)
2005	WW	Morphine and methamphetamine is sewage sludge and WWTP influent; methadone and morphine in aqueous phase of digested sludge	Khan and Ongerth (2005)
2005	WW	*First report of widespread occurrence of an illicit drug in surface water and wastewater* (cocaine and BZE – benzoylecgonine – in WWTP influent and river)	Zuccato et al. (2005)
2005	F	*First implementation of "sewage epidemiology" to reconstruct community-wide drug usage*	Zuccato et al. (2005)
2005	M	Diacetylmorphine on banknotes	Ebejer et al. (2005)

Table 1 (continued)

Year	Aspect	Unique features of study	References
2006	WW	*First study to target a spectrum of illicit drugs and metabolites (in WWTP influents and effluents)*; those not identified in prior studies: norbenzoylecgonine, norcocaine, cocaethylene, 6-acetylmorphine, morphine-3-D-glucuronide, amphetamine, MDA (3,4-methylenedioxyamphetamine), MDEA (3,4-methylenedioxy-N-ethylamphetamine), EDDP (2-ethylidene-1,5-dimethyl-3,3-diphenylpyrrolidine), 11-nor-9-carboxy-9-THC	Castiglioni et al. (2006)
2006	WW	Codeine, dihydrocodeine, hydrocodone, oxycodone, tramadol in WWTP influents and effluents, and surface water	Hummel et al. (2006)
2006	SS	*First report in peer-reviewed literature of an illicit drug in sewage sludge* (amphetamine in sewage sludge)	Kaleta et al. (2006)
2006	F, monit	*First nationwide monitoring in the USA of illicit drugs in sewage*; study by the Office of National Drug Control Policy (ONDCP) targeted about 100 WWTPs across two dozen regions in the USA (results never published)	See Bohannon (2007)
2006	F	*First multi-country monitoring of cocaine in wastewaters to estimate usage*	See UNODC June (2007)
2007	A	*First targeted analysis of ambient air for an illicit drug*; cocaine quantified in particulates from all air samples around Rome and several other Mediterranean locations (also in air samples archived several years prior)	Cecinato and Balducci (2007)
2007	SS	*First report of an illicit drug in biosolids* (methamphetamine in sewage biosolids)	Jones-Lepp and Stevens (2007)
2007	WW	Norcodeine, THC, THC-COOH in WWTP influents and effluents and surface water	Boleda et al. (2007)
2007	M	BZE and heroin on banknotes	Bones et al. (2007b)
2007	R	*First conference devoted to topic of illicit drugs in the environment; led to first published overview of many of the aspects of the topic* (including scientific, technical, social, privacy, ethical, and legal concerns)	EMCDDA (2007), Frost and Griffiths (2008)
2008	DW	*First data on the occurrence and stepwise removal of illicit drugs at a municipal drinking water treatment plant*	Huerta-Fontela et al. (2008a)
2008	WW	Methadone, EDDP, and cocaethylene in surface waters	Zuccato et al. (2008b)

Table 1 (continued)

Year	Aspect	Unique features of study	References
2008	WW	Cocaethylene, LSD (and nor-LSD and 2-oxo-3-hydroxy-LSD), heroin, Δ9-THC (and 11-hydroxy-THC and nor-THC), (R,R)(-)-pseudoephedrine and (1S,2R)(+)-ephedrine hydrochloride in WWTP influents and effluents	Postigo et al. (2008b)
2008	F, monit	Weekly temporal wastewater fluctuations in various drug classes	Zuccato et al. (2008a)
2008	F	First use of the term "sewage epidemiology" in peer-reviewed literature; perhaps first mentioned in a 2007 interview by Fanelli (Bohannon 2007)	Zuccato et al. (2008a)
2008	F	Creatinine in urine first assessed as means of normalizing drug concentrations across WWTPs (and therefore to facilitate drug usage comparisons across communities); creatinine first analyzed in sewage. Creatinine first proposed as a means for normalizing data by Daughton (2001c)	Chiaia et al. (2008)
2008	WW, monit	First systematic survey of illicit drugs in surface waters	Zuccato et al. (2008b)
2008	M, R	First overview of an illicit drug (cocaine) from banknotes from multiple countries	Armenta and de la Guardia (2008)
2008–2009	R	First major overviews of illicit drugs in the environment	Kasprzyk-Hordern et al. (2009a), Postigo et al. (2008a), Zuccato and Castiglioni (2009), Zuccato et al. (2008a)
2008–2009	R	First major overviews of the analytical approaches used for illicit drugs in the environment	Castiglioni et al. (2008), Postigo et al. (2008a), Zuccato and Castiglioni (2009)
2008–2009	R, M	First major overviews of the analytical approaches used for illicit drugs on money	Armenta and de la Guardia (2008)
2008–2009	EF	First studies regarding the sorption of illicit drugs to sediments, soils, and sewage sludge	Barron et al. (2009), Stein et al. (2008), Wick et al. (2009)
2009	DW	First data on the occurrence and stepwise removal of cannabinoids at a municipal drinking water treatment plant	Boleda et al. (2009)
2009	R	First major overview of illicit drugs in airborne particulates	Postigo et al. (2009)
2009	WW	Ecgonine methyl ester (EME) in WWTP influents; EME possibly in surface water	van Nuijs et al. (2009a), Vazquez-Roig et al. (2010)

Table 1 (continued)

Year	Aspect	Unique features of study	References
2009	WW	*First time that illicit drugs (cocaine, BZE, and morphine) monitored monthly in the sewage from an entire city over the course of a year*	Mari et al. (2009)
2009	sw	*Sweat first proposed as a means of general transfer of drugs not just to sewage (via bathing and laundry) but also to any object in the surrounding environment contacted by skin* (dermal transfer)	Daughton and Ruhoy (2009)
2009	monit	*First geographic spatial surveys; 24-h composite WWTP influent samples representing 65% of population of State of Oregon analyzed for BZE, methamphetamine, and MDMA, and Belgium-wide survey of cocaine, BZE, and ecgonine methylester*	Banta-Green et al. (2009), van Nuijs et al. (2009b, c)
2009	A	*First qualitative report of cannabinols in ambient air aerosols (in Rome)*	Cecinato et al. (2009b)
2009	A	Δ9-Tetrahydrocannabinol, cannabidiol, and cannabinol identified in ambient air particulates	Balducci et al. (2009)
2009	A, monit	*First quantitative study of cocaine in ambient air across several continents*	Cecinato et al. (2009a)
2009	WW	Cannabinoids in surface waters	Boleda et al. (2009)
2010	WW	nor-LSD, O-H-LSD, THC-COOH, OH-THC identified in surface waters (river)	Postigo et al. (2010)
2010	A	*First use of existing air quality monitoring sites for detection of multiple drugs of abuse, including amphetamines, cannabinoids, cocainics, lysergics, and opioids* (Spain)	Viana et al. (2010)
2010	WW	*First enantiomeric speciation analysis of illicit drugs in wastewater; including amphetamines, ephedrines, and venlafaxine*	Kasprzyk-Hordern et al. (2010)
2010	WW	*First identification of buprenorphine in sewage, with concentrations ranging up to 20 ng/L in WWTP influents* (France)	Karolak et al. (2010)
2010	WW	*First survey of wastewaters from US pharmaceutical manufacturing facilities reveals relatively high levels (sub-mg/L) of a range of drugs of abuse: butalbital, carisoprodol, methadone, and oxycodone*	Phillips et al. (2010)
2010	WW	*Comprehensive review of FEUDS*	van Nuijs et al. (2010 – in press)

A=air; DW=drinking water; EF=environmental fate; F=forensics; M=money (banknotes); monit=monitoring; R=review; SS=sewage sludge (and biosolids); sw=sweat; WW=wastewater

2 What Is an "Illicit" Drug?

Any discussion regarding illicit drugs can become confused by the ambiguity as to what exactly defines an illicit drug. Confusion stems from the fact that illicit drugs are not limited exclusively to illegal drugs – that is, drugs with no medical use. Illicit drugs can include active ingredients from bona fide registered pharmaceuticals having valuable therapeutic uses – two common examples being morphine and oxycodone. They can also include active ingredients that are banned from all use under various international conventions or national law, as they are deemed as having no use in health care. Whether a drug is illicit (or illegal) can be dictated by a number of different characteristics, including the chemical structure of the active ingredient or the way in which the drug is manufactured, formulated, labeled, distributed, acquired, or used. Some further discussion is presented below to better describe the circumstances under which a drug is considered "illicit."

2.1 Terminology

There is no single, widely used term that accurately captures the myriad numbers of substances that become abused by habitual or addictive use. The term "illicit drug," while widely used, is not accurate in the sense that most of the widely known abused drugs have bona fide medical uses as licit pharmaceuticals; the few that do not are incorporated in the listings of controlled substances maintained by various countries, such as Schedule I in the USA.

A variety of terms are loosely used – often interchangeably – in discussions regarding illicit drugs. Major terms include street drugs, designer drugs, club drugs, drugs of abuse, recreational drugs, clandestinely produced drugs, and hard and soft drugs. The term "research chemicals" had been used by the clandestine laboratory community as an alternative term for designer drugs – with the original intent being that the chemicals were for legitimate research purposes rather than human use (and therefore not subject to regulation); more recently, however, the manufacture of drug analogs as "research chemicals" has become a gray area of the law and is the bona fide trade of those commercial laboratories synthesizing them for biomedical research. The term "designer" drug was first applied in the 1980s to various analogs of fentanyl and then gained popularity when 3,4-methylenedioxymethamphetamine (MDMA, ecstasy) was introduced to the black market; but, perhaps the most notable first "designer" drugs were introduced in the 1920s (i.e., dibenzoylmorphine and acetylpropionylmorphine). A short history of designer drugs is presented by Freye (2009).

Rather surprisingly, no single illicit drug term exists for capturing the full scope of intended meaning. Regardless of the terminology, much overlap exists with licit pharmaceuticals (those with approved medical uses). This can lead to much confusion or ambiguity as to exactly what the scope of the topic is. The confusion surrounding illicit drug terminology is discussed in papers authored by Sussman and Huver (2006) and Sussman and Ames (2008).

In the overview provided herein regarding the environmental aspects of illicit drugs, the guiding definition used is that of the United Nations Office on Drugs and Crime (UNODC), which focuses not on the chemical identity of the drug itself, but rather on the life cycle pathway traveled by a drug. The UNODC does not recognize any distinction between the chemical identity of licit and illicit drugs – only the way in which they are used (UNODC 2009b). In this sense, the term "illicit" refers to the way in which these drugs are manufactured, formulated, distributed, acquired, and consumed and by the fact that they are being used for non-medical purposes – that is, obtaining drugs without a bona fide prescription and using them in the absence of medical supervision.

This definition allows the inclusion of legal pharmaceuticals – that is, when they are manufactured, formulated, distributed, trafficked, or used illegally or diverted from legal sources. For those illicit drugs that originate from diversion of legitimate pharmaceuticals, the many sources and the means for their control to reduce their entry to the environment have been discussed by Ruhoy and Daughton (2008). For those that have illegal origins, the sources and routes to the environment are illustrated in Fig. 1. The wide spectrum of sources, and the routes by which legal drugs become diverted for illicit use, range from the relatively large-scale diversion from pharmaceutical manufacturers, distributors, pharmacies, and health-care facilities to the smaller scale (e.g., "theft" from home storage locations for teen "pharming") and reuse of used medical devices, especially transdermal medical patches, which present lethal hazards for both intentional and accidental exposures (Daughton and Ruhoy 2009).

A closely allied aspect of illegal drugs is counterfeiting. Counterfeiting may involve the repackaging of medical pharmaceuticals that have been either diverted from legitimate sources or manufactured illegally, or the substitution of the advertised ingredient with other substances. Counterfeit is therefore not necessarily synonymous with "fake." Counterfeiting can involve the addition of adulterants to the legitimate pharmaceutical, substitution with less-costly but illegally acquired active pharmaceutical ingredients, or substitution with potentially toxic non-pharmacologic substances. Counterfeit drugs are recognized as a significant threat to human health as a result of the presence of an undeclared active ingredient, excessive dose of a declared ingredient, or absence of a declared active ingredient (WHO 2008). Counterfeiting results in the entry of drugs to legal and illegal distribution channels; drugs can pretend to be either illicit or legitimate. The actual scope of counterfeiting worldwide is not known, but available data indicate it to be enormous and escalating. Of the pharmaceuticals in the developed world, one estimate is that 1% are counterfeit, and in the developing world 10–50% may be counterfeit (Everts 2010). Although counterfeiting often produces drug ingredients that are illegal, it is excluded from the scope of the discussion here.

The scope of this discussion also includes all other chemicals associated with the illegal manufacture (including reformulation of diverted pharmaceuticals) or trafficking of drugs, such as adulterants and impurities (Table 2). With these distinctions acknowledged, the following discussion will tacitly use a variety of terms very loosely. When the term "pharmaceutical" is used, the intention is to reference

Table 2 Adulterants and impurities in illicit drugs (a very small sampling)

Cocaine	MDMA (ecstasy: 3,4-methylenedioxymethamphetamine)
α- and β-truxillines (probably photodimers of cinnamoyl cocaines)	1-(3,4-Methylenedioxy)phenylpropanol-2
3,4,5-Trimethoxycocaine	1-(1,2-Dimethyl-1-azacyclopropyl)methyl-3, 4- methylenedioxybenzene
Benzoyl pseudotropine	1,2-(Methylenedioxy)-4-methylbenzene
Benzoyltropine	1,2-(Methylenedioxy)-4-(2-*N*-
cis- and *trans*-Cinnamoyl ecgonine (hydrolysis of *cis*- and *trans*-cinnamoyl cocaine)	methyliminopropyl)benzene
	1,2-(Methylenedioxy)-4-propylbenzene
cis- and *trans*-Cinnamoyl cocaine (aka methylecgonine cinnamate) (up to 5% by weight)	1,2-Dimethoxy-4-propenylbenzene
	3,4-Methylenedioxyphenyl-2-propanol (MDP)
	3,4-Methylenedioxy-phenyl-2-propanone (MDP2P)
Cuscohygrine (pyrrolidine alkaloid in coca)	3,4-Methylenedioxyamphetamine (MDA)
Diastereomers of synthetic cocaine (pseudococaine, allococaine, allopseudococaine, D-enantiomer of cocaine)	3,4-Methylenedioxy-*N*-methylbenzylamine (MDB)
	3,4-(Methylenedioxy)benzaldehyde
	4-Methoxy-*N*-dimethyl-benzeneethanamine
Diltiazem (adulterant)	4-Methyl-5-phenyl pyrimidine
Ecgonine methyl ester (hydrolysis of cocaine)	Dextromethorphan (adulterant)
Ecgonine (hydrolysis of cocaine)	Dimenhydrinate (adulterant)
Hydroxytropacocaine	Isosafrole
Methylecgonine	Safrole
N-formyl-cocaine	*N*-formyl-3,4-methylenedioxymethamphetamine (*N*-formyl-MDMA)
Norcocaine	*N*-formyl-amphetamine
Tropocaine	*N*-formyl-methamphetamine
Phenacetin (eup to 50% by weight) (adulterant)	*N*-ethyl-3,4-MDA (MDEA)
Xylazine (adulterant)	*N,N*-dimethyl-MDA
Hydroxyzine (adulterant)	*N*-ethyl-*N*-methyl-(1,2-methylenedioxy)-4-(2-aminopropyl)benzene
Hygrine (pyrrolidine alkaloid in coca)	
Levamisole (up to 4% by weight) (adulterant)	*N,N*-dimethyl-(1,2-methylenedioxy)-4-(2-aminopropyl)benzene
Lidocaine (adulterant)	
	Piperonal

Methamphetamine	Heroin
1-Benzyl-3-methylnaphthalene	(Z)-*N*-acetylanhydronornarceine
1,2-Dimethyl-3-phenylaziridine	6-Acetylmorphine
1,3-Dimethyl-2-phenylnaphthalene	3-*O*,6-*O*,*N*-triacetylmorphine
3,4-Dimethyl-5-phenyloxazolidine	3,6-Dimethoxy-4,5-epoxyphenanthrene
cis-1,2-Dimethyl-3-phenylaziridine	4-*O*-acetylthebaol
cis-3,4-Dimethyl-5-phenyl-2-oxazolidone	4,6-Diacetoxy-3-methoxyphenanthrene
Dimethyl amphetamine	4-*O*-Thebaol
Dimethylsulfone (adulterant)	6-*O*,*N*-Diacetylnorcodeine
N-benzyl amphetamine	(E)-*N*-acetylanhydronornarceine
N-acetyl methamphetamine	Acetylcodeine
N-methyl ephedrine	Meconine
N-methyl pseudoephedrine	Clenbuterol (adulterant)
N-ethyl methamphetamine	*N*-acetylnorlaudanosine

Table 2 (continued)

Methamphetamine	Heroin
N-formyl amphetamine	N-acetylnornarcotine
N-acetyl ephedrine	Noscapine (up to 60% by weight)
N-ethyl amphetamine	Papaverine (up to 20% by weight)
N-formyl methamphetamine	1-Methyl-4-phenyl-1,2,3,6-tetrahydropyridine
N,N-dimethyl amphetamine	(MPTP) [during synthesis of
p-Bromotoluene	1-methyl-4-propionoxypyridine (MPPP), an
Phenyl-2-propanone (P2P)	analog of meperidine]

the active ingredients legally registered for use in drugs consumed for approved medical use under formal medical supervision.

What constitutes an illicit drug is a complicated function of social mores and evidence-based health studies, which are sometimes at odds with one another. These conflicts and inconsistencies are reflected, for example, in the opinions expressed by Nutt (2009), which have served to catalyze increasing scrutiny and debate. Illicit substances (drugs and the precursors used for their manufacture) are captured on various government lists (controlled substance *schedules*) that attempt to control and limit their use. The primary criteria justifying inclusion on such listings are health risks, potential for abuse/addiction (partly based on actual data), therapeutic value, and utility as precursors for illicit manufacturing. The unifying worldwide scheme, used by the EU, for regulation of illicit substances comprises the Schedules of the three UN Conventions of 1961 (United Nations Single Convention on Narcotic Drugs, New York, amended 1972), 1971 (Convention on Psychotropic Substances, Vienna), and 1988 (Convention Against Illicit Traffic in Narcotic Drugs and Psychotropic Substances, introducing control on precursors, Vienna). Combined, these Schedules currently comprise about 250 explicitly named controlled substances, according to the European Monitoring Centre for Drugs and Drug Addiction (EMCDDA 2009b).

The lines of demarcation between licit and illicit drugs have become blurred. To illustrate, prescription analgesic opioids (which are controlled prescription drugs; CPDs) have now superseded heroin and cocaine in the USA in leading to fatal drug overdoses (Paulozzi and Xi 2008). Indeed, the use of certain licit drugs, including over-the-counter (OTC) medications, for non-medical purposes has recently surpassed the use of illicit drugs (NIDA 2008). For example, of the top 10 drugs that are misused by high-school seniors in the USA, 7 were legal prescription or OTC medications. Emergency room visits resulting from prescription opioid analgesics more than doubled from 2004 to 2008 and were highest for oxycodone, hydrocodone, and methadone (Cai et al. 2010).

Numerous other illicit substances (such as structural analogs) exist but can only be captured implicitly by generalized chemical criteria that preemptively ban their synthesis; not all countries, however, implicitly capture chemical analogs in their regulations. For example, the US Analogue Act (21 U.S.C. § 813: http://www.justice.gov/dea/pubs/csa/813.htm) is a section of the US Controlled

Substances Act that specifies "A controlled substance analogue shall, to the extent intended for human consumption, be treated, for the purposes of any Federal law as a controlled substance in schedule I." Many additional substances are produced or used illicitly, but their chemical identities are elucidated only after they have experienced sufficient illegal use (often, once adverse medical problems in the general population are documented). A central reference that provides the chemical structures for many of these substances (those listed by the Canadian Controlled Drugs and Substances Act) is maintained on a web page by Chapman (2009).

Further confusion is added to the distinctions between illicit drugs and medical pharmaceuticals because the laws dealing with illicit drugs vary dramatically from country to country. Long-standing drug policies in certain countries are also in a state of flux, as various changes are being considered or are underway. Such changes range from "reducing harm" (e.g., via decriminalization of possession and use) to acknowledgment from the American Medical Association regarding the medical benefits of a Schedule I drug (i.e., namely cannabis) and calling for its clinical research (AMA 2009). Since Portugal began decriminalizing drug use, possession, and acquisition by drug end-users in 2001 (Law no. 30/2000, which focuses on harm reduction) (Greenwald 2009), the spectrum of laws dealing with illicit drugs has diversified; but, growing, illegal manufacturing, and trafficking remain criminal offenses. Among the EU States, the spectrum of law is captured by EMCDDA (2009a). The approaches and evidence used for classifying drugs as illicit are under increasing evidence-based scrutiny and debate (e.g., see Nutt 2009).

2.2 Differences Between Illicit and Licit Drugs as Environmental Contaminants

The primary factor distinguishing illegal from licit (registered) drugs is that the former have no legal (registered) uses, whereas the latter may experience illegal usage. With respect to understanding their overall significance in the environment, seven aspects of illicit drug use contrast sharply with legitimate pharmaceutical use:

(1) For most illicit drugs, there are no accurate quantitative data available on their production or usage. For regulated pharmaceuticals, sales figures and regional real-time prescription data can be used in models to calculate predicted environmental concentrations (PECs); these values can then be compared with measured environmental concentrations (MECs).

(2) Although the chemical identities for the core group of illicit drugs are known, an ever-increasing number of new drugs (such as structural analogs with minor modifications of regulated pharmaceuticals and of previously known illicit drugs – so-called designer drugs or clandestinely produced drugs) can elude detection by forensics laboratories for years before they are noticed and identified. The myriad numbers of designer drugs and constant synthesis of new ones

will pose challenges for mass spectrometrists in the coming years and introduces great uncertainty to the true scope of synthetic chemicals that actually contaminate the environment; for example, see the Psychonaut Web Mapping Research Group (2010) and EMCDDA (2010). Although many of these unique chemicals are probably produced in relatively small quantities, the fact that they belong to relatively few chemical classes may mean that they share relatively few mechanisms of biological action (MOAs). This increases the probability of biological effects resulting from dose (or concentration) "additivity." When multiple chemical toxicants in a mixture share the same MOA, the dose or concentration of each toxicant can add to that of the others. Even if the concentration of each individual toxicant is below an effect threshold, the mixture's combined dose can elicit effects as if it constitutes a single larger dose – a phenomenon informally referred to as "something from nothing" (Kortenkamp et al. 2009). Dose additivity is distinct from potentiation, where a chemical having no biological action of its own can enhance the action of another. Some designer drugs are highly potent, having extremely low effective doses (e.g., in the range of 1 μg per human use), and this has environmental implications, especially for aquatic exposure. As examples, cis-3-methylfentanyl and β-hydroxy-3-methylfentanyl (as with carfentanyl, a large animal tranquilizer) are extraordinarily potent designer drugs – being 3–5 orders of magnitude more potent than morphine.

(3) Drugs manufactured via illicit routes are commonly contaminated with unintended impurities and purposeful adulterants (Table 2). These are often present at extremely high levels (e.g., sometimes more than half of the total mass, as opposed to mg/kg [ppm] levels for impurities in registered medicines) and are often more toxic than the sought-after drug ingredient.

(4) The manufacture of illicit drugs (particularly methamphetamine) can cause extensive ecological damage as well as irreversible damage to infrastructure such as buildings (Cohen et al. 2007; Snell 2001; USEPA 2009a).

(5) The primary interest in residues of illicit drugs in the environment has not been their occurrence in the environment as contaminants, but rather their presence in sewage (mainly untreated raw sewage) for use as a tracking tool to calculate levels of their community-wide consumption. This relatively new tool has been termed *sewage (or sewer) forensics (or epidemiology)*, but later in this chapter is referred to as FEUDS: "Forensic Epidemiology Using Drugs in Sewage." In contrast to the licit use of pharmaceuticals, interest in the potential for illicit drugs as biological stressors in the environment has been secondary, and very little is known.

(6) Compared with pharmaceuticals, much less is known about the toxicology (including pharmacokinetics), especially in the aquatic environment, of many illicit drugs (particularly designer drugs); for human research, there are added legal and ethical difficulties in performing studies on them. Knowledge of the scope of bioactive metabolites and extent of reversible conjugation is comparatively limited.

(7) Numerous measures are routinely implemented to reduce the entry of licit pharmaceuticals into the environment and moderate their potential for adverse effects. Routes of entry span an enormous spectrum of possibilities (Daughton and Ruhoy 2008). With illicit drugs, pollution prevention measures are straight-forward but more difficult to implement – namely, discourage their manufacture, distribution (e.g., via unapproved "rogue" Internet pharmacies), and end use (Fig. 1).

The rate of introduction of new pharmaceuticals with potential for abuse and of new illicit substances precludes any comprehensive definitive worldwide compilation of such chemicals. The INCB (International Narcotics Control Board) maintains three major listings (INCB 2009): Yellow List (Narcotic Drugs under International Control), Green List (Psychotropic Substances under International Control), and Red List (Precursors and Chemicals Frequently Used in the Illicit Manufacture of Narcotic Drugs and Psychotropic Substances under International Control). A convenient listing of many of the corresponding chemical structures is provided by Chapman (2009).

3 The Core Illicit Drugs and the Environment

The types of drugs commonly abused are categorized in various ways, depending on their origin and biological effect. They can either be naturally occurring, semi-synthetic (chemical manipulations, such as analogs, of substances extracted from natural materials), or synthetic (created entirely by laboratory synthesis and manipulation). The primary categories are opiates, other CNS depressants (sedative-hypnotics), CNS stimulants, hallucinogens, and cannabinoids.

The scope of chemicals that could be considered illicit can be viewed in terms of the following categories of medical efficacy:

(1) no known medical use (which are illegal in all circumstances according to various conventions) (e.g., benzylpiperazine; or heroin in the USA),
(2) limited established medical use but also manufactured illegally and used primarily for non-medical purposes (e.g., methamphetamine),
(3) firmly established with wide medical use but diverted for illegal use (e.g., theft; illegal prescription such as via unapproved Internet "pharmacies"),
(4) firmly established wide medical use and legally obtained, but for non-medical use (e.g., doctor/hospital shopping or by other con schemes),
(5) biological action similar to prescription drugs but synthesized as analogs, which are not individually and explicitly categorized as illegal; examples include the numerous analogs of phosphodiesterase (PDE) type-5 inhibitors.

All of these categories tend to primarily comprise drugs with high potential for abuse or recreational use.

Residues of some drugs in the environment have substantial multiple origins (both legal and illegal) making it difficult to ascribe or apportion monitored levels to illicit use. Morphine is one example. Morphine residues can originate from medical use of morphine itself or from codeine (via *O*-demethylation). It can also originate from diverted morphine or codeine as well as from heroin. By collecting data on other (and more unique) metabolites, these pathways can be teased apart. Using morphine as an example, by monitoring for the heroin metabolite 6-AM (6-acetylmorphine), a more reliable idea can be obtained to ascribe what portion of morphine originates from heroin usage.

While drug usage patterns and prevalence vary among countries and with time, those drugs in frequent use in the USA can serve as an organizing framework for further discussion. The annual reports of the US DEA's NFLIS (Drug Enforcement Administration's National Forensic Laboratory Information System) (USDEA 2008) provide the best insights regarding which known drugs are most used in non-medical circumstances (Table 3). The NFLIS is a system operated by the DEA that collects data generated by state and local forensic laboratories in the USA. Of all the samples analyzed in 2008 by US local and state forensic laboratories for the presence of non-medically used drugs, 25 controlled substances composed 90% of all the samples.

Of these 25 drugs, the most frequent 4 were tetrahydrocannabinol (THC), cocaine (benzoylmethylecgonine), methamphetamine, and heroin. Seven were narcotic analgesics (codeine, hydrocodone, oxycodone, methadone, morphine, buprenorphine, and hydromorphone), four were benzodiazepines (alprazolam, clonazepam, diazepam, and lorazepam), and others included 3,4-methylenedioxymethamphetamine (MDMA), 3,4-methylenedioxyamphetamine (MDA), amphetamine, methylphenidate, phencyclidine (PCP), pseudoephedrine, carisoprodol, 1-benzylpiperazine (BZP), and psilocin. In addition to these top 25, other drugs frequently used for non-medical purposes included narcotic analgesics (butorphanol, dihydrocodeine, fentanyl, meperidine, nalbuphine, opium, oxymorphone, pentazocine, propoxyphene, and tramadol), benzodiazepines (chlordiazepoxide, flunitrazepam, midazolam, temazepam, and triazolam), "club" drugs [ketamine, 1-(3-trifluoromethylphenyl)piperazine (TFMPP), gamma-hydroxybutyrate/gamma-butyrolactone (GHB/GBL), 5-methoxy-*N,N*-diisopropyltryptamine (5-MeO-DIPT), and 3,4-methylenedioxy-*N*-ethylamphetamine (MDEA)], a number of stimulants (e.g., cathinone, ephedrine, and phentermine), and a number of anabolic steroids (e.g., methandrostenolone, nandrolone, and stanozolol). Many of these latter drugs (not the top 25) have never been routinely targeted for monitoring as environmental contaminants.

The top 25 detected by NFLIS (DEA's National Forensic Laboratory Information System) are all among the most commonly abused drugs in the USA. The major ones missing from these top 25 (but which are captured in the remaining 10% of samples analyzed by NFLIS) are barbiturates (e.g., phenobarbital and seconal, whose rate of abuse has been declining), certain benzodiazepines (such as alprazolam, chlordiazepoxide, and diazepam, but excepting flunitrazepam), methaqualone, mescaline (3,4,5-trimethoxyphenethylamine), and dextromethorphan (NIDA 2009). Extensive statistics on rates of drug use worldwide (including those maintained by

Table 3 Drugs of abuse frequently detected by US forensics laboratories[a]

Among the 25 abused drugs most frequently detected by US forensics labs	Other abused drugs frequently detected by US forensics labs
Most frequent	*Narcotic analgesics*
Tetrahydrocannabinol (THC)	Butorphanol
Cocaine (benzoylmethylecgonine)	Dihydrocodeine
Methamphetamine	Fentanyl
Heroin (diacetylmorphine; diamorphine)	Meperidine
	Nalbuphine
Narcotic analgesics	Opium
Buprenorphine	Oxymorphone
Codeine	Pentazocine
Hydrocodone	Propoxyphene
Hydromorphone	Tramadol
Methadone	
Morphine	*Benzodiazepines*
Oxycodone	Chlordiazepoxide
	Flunitrazepam
Benzodiazepines	Midazolam
Alprazolam	Temazepam
Clonazepam	Triazolam
Diazepam	
Lorazepam	*"Club" drugs*
	1-(3-Trifluoromethylphenyl)piperazine (TFMPP)
Others	3,4-Methylenedioxy-*N*-ethylamphetamine (MDEA)
1-Benzylpiperazine (BZP)	5-Methoxy-*N,N*-diisopropyltryptamine (5-MeO-DIPT)
3,4-Methylenedioxyamphetamine (MDA)	
3,4-Methylenedioxymethamphetamine (MDMA)	Gamma-hydroxybutyrate/gamma-butyrolactone (GHB/GBL)
Amphetamine	Ketamine
Carisoprodol	
Methylphenidate	*Stimulants*
Phencyclidine (PCP)	Cathinone
Pseudoephedrine	Ephedrine
Psilocin	Phentermine
	Anabolic steroids
	Methandrostenolone
	Nandrolone
	Stanozolol

[a]US DEA's National Forensic Laboratory Information System (USDEA 2008)

the UNODC) can be located from the web page of the Office of National Drug Control Policy (ONDCP 2009). The UNODC World Drug Report (UNODC 2009a) provides comprehensive statistics on world illicit drug supply and demand. The availability, use, and impacts of illicit drugs in the USA were most recently assessed by the National Drug Intelligence Center (NDIC 2010).

3.1 Environmental Occurrence

While drug usage patterns and prevalence vary among countries and through time, those drugs in frequent use in the USA can serve as an organizing framework for further discussion. Of the top 25 most frequently identified, non-medically used, controlled substances analyzed by US local and state forensic laboratories in 2008, only 15 have been targeted in environmental studies of illicit drugs: amphetamine, cocaine, codeine, heroin, hydrocodone, MDA, MDMA, methadone, methamphetamine, methylphenidate, morphine, oxycodone, PCP, pseudoephedrine, and THC (Δ9-tetrahydrocannabinol). A summary of their occurrence in a variety of environmental compartments is shown in Table 4. Note that groundwater is not listed. This is because of the dearth of groundwater monitoring studies that have targeted and identified illicit drugs. One of the only such studies identified codeine in recharged groundwaters in Spain, at sub-ppb levels (Teijon et al. 2010).

Also shown in Table 4 is the occurrence information (as well as indications of negative occurrence – or data of absence) for nearly all of the other illicit drugs and metabolites that have been reported in the published literature. From these data, those analytes with absence of data (i.e., those that have yet to be targeted in monitoring studies) can be deduced. These substances with absence of data represent potential candidates for future monitoring, should they be of interest to environmental scientists, to aquatic toxicologists, or for application with FEUDS. For example, Postigo et al. (2008a) note that nor-cocaethylene and ecgonine ethyl ester have not been targeted in any monitoring study.

The occurrence data are arranged in Table 4 according to the environmental compartments for which the data apply: wastewaters, surface waters, drinking water, sewage sludge, sewage biosolids, air, banknotes, wildlife tissue, and potential for dermal transfer. Dermal transfer is a potential route of transport to immediate physical surroundings (and to sewage during bathing) for drugs excreted via sweat or applied topically (Daughton and Ruhoy 2009). Other reviews of illicit drugs in the environment are provided by Huerta-Fontela et al. (2010) and Zuccato and Castiglioni (2009). It is important to note that parent drugs or their metabolites that have never been targeted for monitoring in the environment are not listed in Table 4. Some of these substances may make likely candidates for future screening. One example is the primary metabolite of methamphetamine, p-hydroxymethamphetamine, which is excreted as the sulfate and glucuronide conjugates (Boles and Wells 2010).

An examination of Table 4 reveals that the drugs with the most positive occurrence data across all environmental compartments are among the top 25 detected by NFLIS – notably the following seven, codeine, morphine, methadone, amphetamine, methamphetamine, cocaine, and THC, and the primary metabolites of methadone (i.e., 2-ethylidene-1,5-dimethyl-3,3-diphenylpyrrolidine [EDDP]), cocaine (i.e., BZE [benzoylecgonine]), and THC (i.e., 11-nor-9-carboxy-9-THC [THC-COOH]). Although widely detected in clinical and forensic drug screens, the occurrence of heroin (diacetylmorphine) in an environmental compartment is limited primarily to banknotes, because of its propensity to hydrolyze in water.

Table 4 Drugs of abuse targeted and identified in environmental compartments[a]

	Wastewaters	Surface waters	Drinking water	Sewage sludge	Biosolids	Air	Banknotes	Wildlife tissue	Dermal transfer[b]
Analgesics									
6-AM (6-acetylmorphine; deacetylated heroin)	×√	×√	×				√		▲
6-AC (6-acetylcodeine)		×							▲
Codeine[c]	√√√	√√	××√				×		▲
Dihydrocodeine[d]	√	√							
Heroin (diacetylmorphine)[c]	××√	×	×	√√		√	√√		▲
Hydrocodone[c]	√√	×√					√		
Morphine[c]	√√√	√	×			×			
Morphine-3β-D-glucuronide	×√	×							
Norcodeine	√	√	××√						
Normorphine	√	×	×						
Fentanyl[d] (excreted mainly as norfentanyl)	××	××	×						‡‡
Norfentanyl	×								
Oxycodone[c]	√√	√							
Tramadol[d]		√						√	
Methadone									
Methadone[c]	√√√	√	√√	√√√		×			▲
EDDP (2-ethylidene-1,5-dimethyl-3,3-diphenylpyrrolidine)	√√√	√√	√√						
Stimulants									
Amphetamine[c]	√√√	×√	××√	√√		×√	√		▲
Ephedrine[d]/pseudoephedrine[c]	√√√	√				×			
Methamphetamine[c]	√	×√	××√		√	×√	√		▲
MDA[c]	√	×√							
MDBD	×								

Table 4 (continued)

	Wastewaters	Surface waters	Drinking water	Sewage sludge	Biosolids	Air	Banknotes	Wildlife tissue	Dermal transfer[b]
MDEA[d]	x√	x							
MDMA[c]	√√	√	√			x√			◄
Methylphenidate[c]									÷
Cocainics									
Benzoylecgonine (BZE)	√√√	√√	√			√	√		◄
Cocaethylene	x√	√√				x			
Cocaine[c]	√√√	√√	x√			√√	√√√		◄◄
Ecgonine methyl ester (EME)	x√	√							
Norbenzoylecgonine	√	√							
Norcocaine	x√	√							◄
"Club" drugs (e.g., dissociative anesthetics)									
Ketamine[d]	xx√	x	x						
Norketamine	x								
PCP (phencyclidine)[c]	x√	x	x				√		
Hallucinogens									
LSD	xx√	x	x			x			
Nor-LSD (*N*-desmethyl-LSD)	x√	xx√				x			
O-H-LSD (2-oxo-3-hydroxy-LSD)	x√	xx√				x			
Cannabinoids									
Cannabinol (CNB)						√√	x√		
Cannabidiol (CND)						x√	x√		
OH-THC (11-hydroxy-Δ9-tetrahydrocannabinol)	x√	xx√				x			
nor-THC						x			
THC (Δ9-tetrahydrocannabinol)[c]	√√	√√	x			√√	x√		◄
THC-COOH (11-nor-carboxy-Δ9-tetrahydrocannabinol)	√√√	√√	x						

Table 4 (continued)

	Wastewaters	Surface waters	Drinking water	Sewage sludge	Biosolids	Air	Banknotes	Wildlife tissue	Dermal transfer[b]
Other									
Flunitrazepam[d]	×								
Testosterone									‡

"×": Frequency of negative occurrence data (data of absence); supporting data are stronger with more "×" (up to two total)

√: Frequency of positive occurrence data; supporting data are stronger with more "√" (up to three total)

Blank cells denote lack of any type of supporting data (absence of data)

Supporting references

Wastewaters: Bartelt-Hunt et al. (2009), Bijlsma et al. (2009), Boleda et al. (2007, 2009), Bones et al. (2007a), Castiglioni et al. (2006, 2007), Chiaia et al. (2008), Frost and Griffiths (2008), Gheorghe et al. (2008), González-Mariño et al. (2009, 2010), Huerta-Fontela et al. (2007, 2008a, b), Hummel et al. (2006), Jones-Lepp et al. (2004), Karolak et al. (2010), Kasprzyk-Hordern et al. (2007, 2008a, b, 2009a, 2010), Khan (2002), Loganathan et al. (2009), Mari et al. (2009), Postigo et al. (2008b, 2010), Terzic et al. (2010), van Nuijs et al. (2009a), Zuccato et al. (2005, 2008a)

Surface waters: Bartelt-Hunt et al. (2009), Bijlsma et al. (2009), Boleda et al. (2007, 2009), Bones et al. (2007a), Gheorghe et al. (2008), González-Mariño et al. (2010), Huerta-Fontela et al. (2007, 2008a), Kasprzyk-Hordern et al. (2007, 2008a), Postigo et al. (2010), Zuccato et al. (2008b, 2005)

Drinking water: Boleda et al. (2009), Huerta-Fontela et al. (2008a)

Sewage sludge: Kaleta et al. (2006), Khan (2002)

Biosolids: Jones-Lepp and Stevens (2007)

Air: Balducci et al. (2009), Cecinato and Balducci (2007), Cecinato et al. (2009a, b, 2010), Hannigan et al. (1998), Postigo et al. (2009), Viana et al. (2010)

Banknotes (small sampling of published works): Aaron and Lewis (1987), Armenta and de la Guardia (2008), Bones et al. (2007b), Carter et al. (2003), Ebejer et al. (2005, 2007), Felix et al. (2008), Jenkins (2001), Lavins et al. (2004), Sleeman et al. (2000), Zuo et al. (2008)

[a]The references providing the data for this table are listed for each of the columns. Wastewaters include both raw sewage influent and treated sewage effluent. Note that this table does not include drugs or metabolites that have never been targeted in monitoring studies, *p*-hydroxymethamphetamine is one example

[b]Potential for transfer from skin to surroundings (Daughton and Ruhoy 2009): ▲ = known to be excreted via sweat; ‡ = available in high-concentration dermal transfer devices

[c]15 of the top 25 most frequently identified, non-medically used, controlled substances – as analyzed and reported by US local and state forensic laboratories in 2008 (see USDEA 2008) and which comprised 90% of all drugs identified (USDEA 2008)

[d]Among the other most frequently identified, non-medically used, controlled substances – as analyzed and reported by US local and state forensic laboratories in 2008 (see USDEA 2008); the other 9 most frequently identified 25 drugs, but not yet targeted in more than a single environmental study, focused expressly on illicit drugs are alprazolam, buprenorphine, BZP (1-benzylpiperazine), carisoprodol, clonazepam, diazepam, hydromorphone, lorazepam, and psilocin

Similarly, the cannabinoids are detected most frequently in air. Not surprisingly, no illicit drug (or metabolite) frequently reported with environmental occurrence data is missing from the 25 most frequently identified by forensic labs.

Nine of the remaining 25 drugs most frequently identified by the forensic testing labs have not yet been targeted in environmental studies whose primary focus is illicit drugs. These are alprazolam, buprenorphine, BZP (1-benzylpiperazine), carisoprodol, clonazepam, diazepam, hydromorphone, lorazepam, and psilocin (4-hydroxy-dimethyltryptamine, 4-HO-DMT). Of these nine drugs, environmental occurrence data have been published in studies targeted at CPDs for alprazolam, carisoprodol, diazepam, and lorazepam. Data do not exist for buprenorphine, BZP, clonazepam, hydromorphone, and psilocin. Depending on their pharmacokinetics and the extent to which that are excreted unchanged, these latter five drugs may be likely targets for future environmental monitoring.

Alprazolam has been measured at low to high ng/L levels in treated sewage effluent (Batt et al. 2008). Although carisoprodol is extensively metabolized (primarily to the active metabolite meprobamate), it has been measured at sub-ppb levels in runoff from agricultural fields irrigated with treated wastewater (Pedersen et al. 2005).

Diazepam has been widely reported in a variety of wastewaters and surface waters; see the summaries of Calisto and Esteves (2009) and Straub (2008). Most diazepam occurrence data from targeted monitoring, however, have been negative (Christensen et al. 2009). Diazepam resists biodegradation (Redshaw et al. 2008) and perhaps partitions to particulates.

Some illicit drug analytes, when targeted, are infrequently reported, possibly as a result of their considerably higher detection limits. Normorphine and THC-COOH are examples, sometimes having limits of detection 1–2 orders of magnitude higher than those of other analytes. This reiterates the importance of specifying limits of detection when presenting data of absence.

Other targeted analytes are not detected, probably because they are extensively metabolized or excreted as conjugates. Conjugation undoubtedly plays a critical role in determining whether a free parent drug will be found in waters. Many drug ingredients are extensively conjugated and, without a hydrolysis step to free the aglycone, will be missed (Daughton and Ruhoy 2009; Pichini et al. 2008). Conjugates could potentially serve as hidden reservoirs for drug ingredients in the environment (Daughton 2004), but, to date, published data are lacking to affirm the extent and magnitude of this phenomenon.

Lorazepam is extensively metabolized to its glucuronide conjugate, with negligible amounts excreted unchanged (Ghasemi and Niazi 2005). Nonetheless, it has been measured at levels up to 200 ng/L in treated sewage (Coetsier et al. 2009; Gros et al. 2009, 2010), perhaps reflecting an input from disposal to sewers or hydrolysis of the conjugate.

It is important to note that some illicit drugs are metabolic/transformation daughter products of others, which explains why their concentrations in sewage or receiving waters are routinely higher than those of their parents. One example is heroin, which is quickly deacetylated (both metabolically and in the environment) to 6-AM followed by hydrolysis to morphine. This means that the probability is higher that these parent drugs, when detected in waters (especially waters removed from impact by sewage), are present because they were directly flushed into sewers (or excreted via sweat) rather than being excreted via urine. An alternative source could be runoff into streams, such as during clandestine manufacturing. Another example is fentanyl, which is extensively excreted as norfentanyl.

3.2 Adulterants and Impurities as Potential Environmental Contaminants

In contrast to pharmaceuticals produced under Good Manufacturing Practices, drugs made illegally contain significant impurities and contaminants in addition to the sought-after drug (or sometimes even in place of the desired drug). These substances are often present at very high levels, especially in intentionally mislabeled drugs – sometimes representing the bulk of the purported drug. For example, noscapine can be present at levels up to 60% in heroin, or phenacetin at levels up to 50% in cocaine. Another example is the misrepresentation of MDMA by combining 1-benzylpiperazine (BZP) and 1-(3-trifluoromethylphenyl)piperazine (TFMPP), which can mimic its psychoactive effects. These adulterants and other contaminants also include products of synthesis or processing (precursors, intermediates, by-products), natural impurities (e.g., natural product alkaloids), products of degradation (e.g., oxidation during storage), and pharmacologically active adulterants (e.g., many licit drugs and other chemicals, such as levamisole, xylazine, lidocaine, phenacetin, hydroxyzine, and diltiazem). Some of these impurities or adulterants are more potent than the sought-after drug (cocaethylene being one example – a synthesis by-product and metabolite of cocaine when consumed together with ethanol). In the course of reviewing the literature, more than 90 common adulterants and impurities were noted just for the four illicit drugs cocaine, MDMA, methamphetamine, and heroin (Table 2). These represent only a small sampling of the variety of chemicals that can compose illicit drugs.

Because some illicit drugs are natural products, they can inadvertently contaminate our food supply. The recent controversy regarding the presence of cocaine in a commercial energy drink (as residue from de-cocainized extract of coca leaf) (BfR 2009) demonstrates the power of analytical chemistry in revealing previously undetected levels of chemicals.

Adulterants are often used to enhance desired biological effects or make the drug more profitable. They include diluents, which are added to mimic the appearance of the sought-after drug (to extend the doses per mass) or enhance the biological effects. Impurities are sometimes integral to the natural chemistry of the native plant from which a drug is isolated and at other times is a function of the synthetic route to the desired drug. The adulterants used are a function of the geographic locale of manufacture/distribution or depend on what chemicals are available at the time of synthesis or what the clandestine manufacturer wishes to use. Many dozens of impurities and adulterants are possible for any given drug synthesis. Impurities, in turn, can each yield numerous metabolites, most of which are known. Adulterants can range from common substances such as caffeine (very high concentrations) to more insidious chemicals such as the cytotoxic veterinary dewormer drug lev-amisole, which has led to a number of deaths from its inadvertent consumption. In this way, illicit drug use can serve as an alternative route of entry to the environment not just for drugs of abuse, but also for active pharmaceutical ingredients, such as levamisole, that have no potential for abuse. Adulteration of illicit drugs has grown to become a major health risk for drug users.

An expansive published literature exists for illicit drug adulterants and impurities. This is driven largely by research and surveillance aimed at drug "profiling," a methodology for obtaining a chemical fingerprint or signature for individual batches of drugs. For example, determining illicit drug impurities (and ratios of enantiomers) helps deduce the synthetic route or geographic locale of manufacture. An example of the profiling process (for methamphetamine) is presented by Inoue et al. (2008). Profiling data are potentially useful for targeting important adulterants or impurities for environmental monitoring.

Except for some registered pharmaceuticals that are used as adulterants in illicit drugs (to reduce cost or alter/mimic physiologic/psychotropic effects), these adulterants pose totally unknown risks for the environment. The ecological risks for some registered pharmaceuticals used as adulterants are similarly unknown. One example is levamisole, which is excreted largely unchanged and potentially poses risks for certain soil-dwelling organisms (McKellar 1997; Sommer and Bibby 2002). It is also known to be taken up by certain food crops such as lettuce (Boxall et al. 2006a), but has not yet been targeted in any environmental monitoring. Levamisole has, however, been identified as a high-priority compound for possible future environmental monitoring (Boxall et al. 2006b).

The general public may be unknowingly exposed to illicit drugs in the form of designer drugs as impurities in food or nutritional supplements. For example, common foods may contain residues of illegal analogs of legal drugs, particularly anabolic hormones (used in livestock), such as norbolethone, tetrahydrogestrinone, and desoxymethyltestosterone (Cunningham et al. 2009; Noppe et al. 2008; Shao et al. 2009; Yang et al. 2009). Certain OTC supplements used for male erectile dysfunction may contain unregistered synthetic analogs of the approved phospho-diesterase type-5 (PDE-5) inhibitors (Poon et al. 2007; Venhuis and de Kaste 2008; Venhuis et al. 2007).

4 Large-Scale Exposure or Source Assessments via Dose Reconstruction

Interest in illicit drugs in the environment has both prospective and retrospective dimensions. The prospective dimension is concerned with the exposure of aquatic organisms and humans to environmental residues. Of the environmental studies conducted, however, this has not been the major thrust. Rather, the data obtained have been used as a retrospective tool for reconstructing society-wide usage of illicit drugs. Such data acquisition could be considered a large-scale version of exposure assessment called "dose reconstruction" (e.g., see ATSDR 2009).

Dose reconstruction approaches that use the presence of drug residues on banknote currency and in airborne particulates have also been attempted. These could be more accurately referred to not as dose reconstruction, however, but rather as source reconstruction (deciphering the source and intensity of the origin of the drugs).

4.1 Sewage Epidemiology or Forensics – FEUDS

Daughton (2001c) first proposed analyzing sewage for residues of illicit drugs unique to actual consumption (rather than originating from disposal or manufacture) for the purpose of back-calculating estimates of community-wide usage rates. Since 2001, this approach has been referred to as "sewage epidemiology" (a term first reported in the literature by Zuccato et al. 2008a), "sewage forensics," and "community-wide urinalysis" or "community drug testing." None of these terms, however, fully captures the multiple purposes that could potentially be served by application of the methodology.

Epidemiology can be defined as the study of the occurrence, distribution, and causes of health effects in specific human populations and the use of this study as the basis for interventions targeted at reestablishing public health. Epidemiology is used for identifying at-risk subpopulations, monitoring the incidence of exposure/disease, and detecting/controlling epidemics. Elements of illicit drug use fit all of these categories. In its simplest state, "forensics" involves the extraction of pertinent information to support an argument or investigation (Daughton 2001b). One of its best known modern applications is to assist in resolving legal issues, and the worldwide legal system plays an integral role in all aspects of illicit drug use.

Since this still-evolving approach for measuring drugs in sewage to estimate collective drug usage has elements of both forensics and epidemiology, it would be more accurately captured under the newer term "Forensic Epidemiology," which integrates the principles and methods used in public health epidemiology with those used in forensic sciences (Goodman et al. 2003; Loue 2010).

Therefore, a more accurate descriptive term for "sewer epidemiology" should be considered to better unify the published literature. One possibility could be "Forensic Epidemiology Using Drugs in Sewage" (FEUDS). Use of a unique term and acronym would have the added benefit of more easily facilitating communication across fields and to greatly simplify literature searches. The acronym FEUDS will be used as a shorthand in the remainder of the discussion here.

4.2 FEUDS for Community-Wide Dose Reconstruction of Illicit Drugs

After its conceptualization in 2001 (Daughton 2001c, d), FEUDS was first implemented in a 2005 field monitoring study by Zuccato et al. (2005). FEUDS was originally proposed as the first evidence-based approach for measuring drug use because the long-practiced approaches that use oral or written population surveys are fraught with limitations, not the least of which involve numerous sources of potential error that are difficult to define, control, or measure (especially sampling bias and self-reporting bias) (Daughton 2001c). The limitations imposed by self-reporting bias have been corroborated in "concordance" studies (comparisons of self-report data with empirical bioanalysis data), which point to gross underreporting by self-reports (often at rates as low as one-half of actual); the problems with profound underestimates derived from self-reporting are discussed by Magura (2010). Sampling bias inevitably results from the decision process used for selecting which segments of the general population to survey.

These conventional approaches to estimating illicit drug usage also suffer from two inherent limitations: extreme delays in time before results are compiled and reported and costs associated with data collection and interpretation.

FEUDS, like public surveys, suffers from many sources of potential error. But FEUDS is in its infancy and its sources of error derive from variables still under investigation and which have not yet been optimized for better control. While conceptually rather straightforward, the back-calculations used in FEUDS are a function of numerous variables, including demographics, population flows through a locale (such as transient visitors and commuters) served by a given sewage treatment facility, route of dose administration, pharmacokinetics (including knowledge of extent of conjugation), constancy of usage, frequency of disposal (if the parent drug rather than a unique metabolite is targeted), and sewage flows. Combined, these pose a major challenge for modeling to accurately reconstruct dose. The numerous problems facing FEUDS are discussed in Frost and Griffiths (2008) and in van Nuijs et al. (2010 – in press). Most FEUDS investigators couple drug concentrations in sewage with per-capita sewage flows to calculate what is sometimes called "index loads" or "per-capita loads," expressed as mg/person/day. Many of the sources of uncertainty are covered by Banta-Green et al. (2009) and Zuccato et al. (2008a).

Despite the plethora of uncertainties attendant to variables involved in back-calculations, the ability to provide estimates of near-real-time community-wide usage is something that is not possible with any other known approach. This also opens the possibility of detecting real-time trends or changes in drug use. Example applications include verifying reductions in drug use as a result of interdictions or public health campaigns or detecting the emergence of newly available drugs or overall changes in drug-use patterns. Data on real-time usage could better inform decisions regarding drug control and mitigation. Correlating policy actions with resulting society-wide impacts cannot be effectively done when collected data are significantly delayed in reporting. Transient or episodic patterns are obscured when reports are on an annual basis.

Few systematic approaches to cataloging newly emerging recreational drugs (those not yet recognized in the published literature) have existed. One such attempt, conducted from 2008 through 2009, mined information collected from a broad spectrum of sources (Psychonaut Web Mapping Research Group 2010). As of March 2010, the project had categorized over 400 substances or mixtures not previously recognized in the published literature as having recreational use. One example is mephedrone (2-methylamino-1-*p*-tolylpropan-1-one, 4-methylmethcathinone, 4-MMC, MMCAT), a substance that has experienced wide and growing popularity as a street drug in the UK but which is sold in various guises, such as "plant food" and labeled "not for human consumption." By mid-April 2010, mephedrone had been banned in the UK, only to witness another drug enter the spotlight – 5,6-methylenedioxy-2-aminoindane (MDAI) – developed in the 1990s as an antidepressant. This exemplifies the speed at which a continual series of new chemicals is embraced by recreational drug users.

It is of great potential significance that there are no apparent technical obstacles to designing automated continuous monitors for use in sewage collection/distribution systems. Implementing continuous monitoring to support FEUDS could greatly enhance efforts to control and mitigate drug use. Such a hypothetical system could use a number of different approaches, generally based on the use of in-stream chemical sensors or automatic acquisition of discrete samples at pre-selected intervals followed by instrumented auto-analysis. The limiting factor would be cost. The foundation for continuous monitoring is already being established, especially for use in clinical and forensics laboratories. One such automated method has been applied to 21 commonly abused drugs in urine, using online extraction coupled with tandem mass spectrometry (Chiuminatto et al. 2010); the main area of needed improvement is sufficiently low limits of detection.

Another advantage of FEUDS over population surveys is that not all drug use is necessarily known to the users themselves, who then unintentionally report incorrect drug identities and usage quantities. Illicit drug users often do not know the identity or the quantity of the active substances they have consumed because the purity of what they consume is unknown. Often, the active substance or quantity is not what the distributor claims (e.g., counterfeit illicit drugs). Adulterants are often substituted (Table 2), in part or in whole, for the purported drug. One general route of such uninformed exposure is the surreptitious incorporation of designer drugs into otherwise legal OTC diet supplements or recreational or lifestyle products. An example is the relatively new (and probably still incompletely characterized) synthetic analogs of the approved phosphodiesterase type-5 (PDE-5) inhibitors (used primarily in treating erectile dysfunction), such as sildenafil, vardenafil, and tadalafil (Poon et al. 2007; Venhuis and de Kaste 2008; Venhuis et al. 2007). In more than half of the OTC male erectile dysfunction health products examined, analyses revealed the presence of acetildenafil, hydroxyacetildenafil, hydroxyhomosildenafil, and piperidenafil – analogs of sildenafil and vardenafil not registered for pharmacologic use. The legal registered versions of PDE-5 inhibitors have only recently been detected in wastewaters (Nieto et al. 2010). Since members of this class of drugs all share the same mechanism of biological action, the PDE-5 inhibitor analogs could contribute to dose additivity. Analogs are known to exist for various other classes of drugs, particularly psychoactives, anabolic steroids, and anti-obesity drugs. The toxicity of these analogs is largely unknown. The extent of such adulteration in the drug and supplements industry is unknown, largely because the targets for analysis are often not known to forensic analysts.

Hagerman (2008) provides a brief history of FEUDS projects in the USA. The ONDCP performed the first FEUDS monitoring in the USA in 2006, targeting about 100 wastewater treatment plants (WWTPs) across two dozen regions (Bohannon 2007). The first conference devoted to FEUDS was organized by EMCDDA in Lisbon, Portugal, in April 2007 (EMCDDA 2007). It led to the first published overview of many of the aspects of the topic (including scientific, technical, social, privacy, ethical, and legal concerns), as provided by Frost and Griffiths (2008).

4.3 Quality Assurance and FEUDS

Two aspects of illicit drugs may have a major impact on the quality and validity of any monitoring data used for FEUDS. The first is the contamination of samples during collection or analysis by transfer of residues from the skin of the analyst. Many drugs, especially illicit drugs, are readily excreted via sweat glands, including those on the fingers. This has the potential to result in contamination of samples during their collection or during various steps in analysis. Contamination of samples by analysts who are using prescribed or illicit drugs is an under-investigated potential source of erroneous data. The dermal excretion of drugs as a source of their transfer to immediate surroundings as well as to the environment was first examined by Daughton and Ruhoy (2009).

The second aspect is the stability of drug residues in samples in the absence of proper preservation. Little research has been done on the stability of illicit drugs in collected environmental samples; the extensive existing literature on the stability of residues in biological samples obtained for forensics and human drug monitoring purposes may be partly relevant and could serve as a starting point for environmental samples. Both cocaine and cocaethylene, for example, have been shown to readily degrade to benzoylecgonine (Castiglioni et al. 2006). González-Mariño et al. (2010) examined the preservation of raw sewage samples with sodium azide at 4°C to inhibit microbial degradation of labile analytes such as cocaine and cocaethylene. In time-course studies up to 7 days, large positive or negative changes in concentrations were noted for methadone, cocaine, benzoylecgonine, heroin, morphine, and THC-COOH. They concluded that sample preparation (e.g., solid phase extraction followed by any needed derivatization and storage at low temperature) was best performed as soon as possible at the site of sample collection.

4.4 Summary of Published Research in FEUDS

Overviews and discussion of the FEUDS studies published up until 2008 are provided by Postigo et al. (2008a), van Nuijs et al. (2010 – in press), and Zuccato et al. (2008a). The major published articles regarding one or more aspects of the FEUDS approach are compiled in the chronology of Table 5. At the beginning of 2010, there had been fewer than two dozen studies, and most were published after 2007.

Table 5 Major FEUDS studies (arranged according to chronology)

Year	Title (citation)
2001	Illicit drugs in municipal sewage: proposed new non-intrusive tool to heighten public awareness of societal use of illicit/abused drugs and their potential for ecological consequence (Daughton 2001c)
	Commentary on illicit drugs in the environment: a tool for public education – societal drug abuse and its aiding of terrorism (Daughton 2001d)
2005	Cocaine in surface waters: new evidence-based tool to monitor community drug abuse (Zuccato et al. 2005)
2006	High cocaine use in Europe and US proven Stunning data for European Countries: first ever comparative multi-country study of cocaine use by a new measurement technique (Sörgel 2006)
2007	Using environmental analytical data to estimate levels of community consumption of illicit drugs and abused pharmaceuticals (Bones et al. 2007a)
2008	Occurrence of psychoactive stimulatory drugs in wastewaters in north-eastern Spain (Huerta-Fontela et al. 2008b)
	Estimating community drug abuse by wastewater analysis (Zuccato et al. 2008a)
	Assessing illicit drugs in wastewater: potential and limitations of a new monitoring approach (Frost and Griffiths 2008)
2009	Cocaine and metabolites in waste and surface water across Belgium (van Nuijs et al. 2009b)
	Cocaine and heroin in wastewater plants: a 1-year study in the city of Florence, Italy (Mari et al. 2009)
	Monitoring of opiates, cannabinoids, and their metabolites in wastewater, surface water, and finished water in Catalonia, Spain (Boleda et al. 2009)
	Can cocaine use be evaluated through analysis of wastewater? A nationwide approach conducted in Belgium (van Nuijs et al. 2009c)
	Illicit drugs and pharmaceuticals in the environment – forensic applications of environmental data, Part 1: estimation of the usage of drugs in local communities (Kasprzyk-Hordern et al. 2009b)
	Municipal sewage as a source of current information on psychoactive substances used in urban communities (Wiergowski et al. 2009)
	The spatial epidemiology of cocaine, methamphetamine, and 3,4-methylenedioxymethamphetamine (MDMA) use: a demonstration using a population measure of community drug load derived from municipal wastewater (Banta-Green et al. 2009)
2010	Drugs of abuse and their metabolites in the Ebro River basin: occurrence in sewage and surface water, sewage treatment plants removal efficiency and collective drug usage estimation (Postigo et al. 2010)
	Estimation of illicit drugs consumption by wastewater analysis in Paris area (France) (Karolak et al. 2010)
	Illicit drugs in wastewater of the city of Zagreb (Croatia) – estimation of drug abuse in a transition country (Terzic et al. 2010)
	Illicit drug consumption estimations derived from wastewater analysis: a critical review (van Nuijs et al. 2010 – in press)

Published FEUDS analyses have been conducted in a number of countries, with assessments at local, regional, or national levels – primarily in Belgium, Germany, Ireland, Italy, Spain, Switzerland, the USA (i.e., Oregon), and Wales. To date, FEUDS assessments have been focused on a select few parent drugs (primarily cannabis, cocaine, heroin, and MDMA) using various metabolites. They have been performed using many sampling methodologies – ranging from 1-day single-event discrete grab sampling to longer term (e.g., 12-month) integrative continuous sampling over numerous WWTPs or rivers, servicing regions with populations exceeding millions. In many of these studies, temporal usage patterns were investigated, in which yearly seasons or the day of the week (e.g., higher cocaine use on weekends) was examined. Usage rates are reported on various comparative bases, often involving per capita (e.g., g/day/1,000 population – usually ranging only up to several grams), total consumption (e.g., tonne per year per geographic area), or flows (mass/river/day). Discrete monitoring must acknowledge the cyclic or episodic drug-use pattern fluctuations in concentrations that can result from diurnal cycles, seasons, or day of the week. This can be particularly pronounced for recreational drugs.

An enormous published literature surrounds the forensic chemistry of illicit drugs. The numbers of illicit drugs analyzed in the environment, however, is a small fraction of those that have been targeted in countless studies published on biological tissues and fluids for the purposes of forensics and patient compliance monitoring and for the study of pharmacokinetics in animals. Accurate-mass (exact-mass) identification of unknowns (e.g., via time-of-flight mass spectrometry – TOF-MS) plays a central role especially when authentic reference standards are not available. While this conventional forensics literature can serve as a guide for environmental analysis, it is only indirectly relevant. There are numerous variables involved with (and impacting) the procedural steps used in the analyses required by FEUDS – ranging from sampling design and matrix interferences to analyte determination and the need for extremely low limits of detection. Some major overviews and discussion of the analytical approaches for measuring illicit drugs in wastewaters and other waters are available (Castiglioni et al. 2008; Postigo et al. 2008a; Zuccato and Castiglioni 2009).

With interest in trace environmental contaminants (or micro-constituents) continuing to grow, a critical and limiting factor in gaining a comprehensive and accurate picture is the limit of detection (LOD) – and allied figures of merit such as the limit of quantitation (LOQ). LOD and LOQ are functions of the individual analyte as well as the matrix in which it occurs; raw sewage, for example, is a particularly problematic matrix, giving significantly higher LODs than drinking water. As a key figure of merit, the LOD dictates the extent to which environmental monitoring produces meaningful data of absence (negative occurrence data); it is roughly defined as the lowest concentration that an analytical method can differentiate with statistical power from background signal. With discussions of the formal definition of the LOD aside, one ramification is that LODs can differ widely among analytes (and among methods). Therefore, data of absence cannot be directly inter-compared without providing the context of their respective magnitudes. The absence of two drugs in a sample, for example, has different meanings when their LODs differ by 1, 2, or even more orders of magnitude. To state that a drug is not found in a certain sample is rather meaningless without specifying its LOD. For most of the monitoring studies cited in this chapter, LODs were provided as part of the method development. For illicit drugs in sewage, LODs tend to settle in the 1–10 ng/L range, with excursions to either side. Some drugs have higher LODs – possibly a reason for sporadic occurrence data. One example is 6-acetylmorphine, whose LOD can be an order of magnitude higher than for others, such as cocaine and cocaethylene (Postigo et al. 2008b).

An issue little addressed in FEUDS studies has been the complications (and opportunities) posed by chirality. Only recently has attention begun to be directed to the speciation of enantiomers during environmental analysis (Kasprzyk-Hordern et al. 2010). Possibly the majority of illicit drugs have at least one chiral center (Smith 2009). The alkaloid truxilline, as an example, occurs in coca leaf as 11 stereoisomers. Amphetamines can each have a pair of enantiomers, sometimes distinguishing the licit from the illicit form (as well as portending relative toxicity). This may account for a portion of some of the large variance in estimated amphetamine usage across FEUDS studies. While chiral isomers can pose difficult challenges for analytical chemists, they also provide a wealth of forensics information in terms of chemical "fingerprinting" – for example, in distinguishing legal from illegal origins. Advancements in the application of chiral analysis to illicit drugs in the environment will most likely accelerate, especially in its use for FEUDS.

4.5 Legal Concerns Surrounding FEUDS

Application of FEUDS to analysis of co-mingled sewage (such as at a sewage treatment facility) clearly ensures the anonymity of individuals, which was one of its primary features when first proposed (Daughton 2001c, d). Even though FEUDS was conceptualized for public health purposes, the potential for its abuse in law enforcement was recognized early. An obvious scenario where privacy could be breached would be the implementation of sewage monitoring as close to individual sewer feeder lines as possible to trace the origin of illicit drug residues back to specific, individual neighborhoods or isolated buildings. Despite this tacit understanding as far back as 2001, there has been little formal discussion of legal or ethical issues in the published literature, even in law journals; interest in more specific, localized application of FEUDS is evident from statements such as whether it "can be used in smaller communities in which illicit drug use is especially unwanted such as drug rehabilitation centers, hospitals, prisons, military compounds and schools" (Verster 2010). One of the only, and certainly the most comprehensive, examinations of the legal concerns (in the USA) was published by Hering (2009). The concerns center primarily on the Fourth Amendment (unreasonable searches) and the potential for violating an individual's privacy. Although the historical summary of events behind FEUDS is not fully accurate, Hering presents a comprehensive examination of the pitfalls involving US law, using case law to substantiate the concerns. He concludes, however, that although FEUDS applied to the sewers of an isolated home might appear to constitute a search under the Fourth Amendment, the legal case would be "extremely tenuous."

5 Illicit Drugs in the Money Supply

Residues of illicit drugs have been known since the 1980s to occur on banknotes (e.g., Aaron and Lewis 1987; Table 1), primarily as a result of dermal transfer from drug users and transfer from contact with bulk drugs themselves. Highly contaminated banknotes can, in turn, cross-contaminate pristine banknotes in their proximity. Most research has been focused on cocaine, because of its propensity to become entrapped in banknote fibers and because of the use of banknotes for insufflation. Cocaine amounts exceeding 1 mg per banknote have been reported (Oyler et al. 1996), more than 1% of a typical dose. The contamination may be so pervasive that large numbers of banknotes must be removed from general circulation each year (Thompson 2002). Bones et al. (2007b) pushed the limit of detection for cocaine into the range of a picogram per banknote. In addition to cocaine, other drugs studied on banknotes include 6-AM, diacetylmorphine (DAM), Δ9-tetrahydrocannabinol, cannabinol, cannabidiol, 3,4-methylenedioxymethamphetamine, methamphetamine, amphetamine, PCP, and codeine.

Although the occurrence of illicit drugs on money in general circulation possibly serves as a minor source of exposure for the public, via dermal transfer and pulmonary exposure (but especially among those working with money sorting machines), no exposure work has been done on these routes. Interest has been spurred instead by forensics – primarily with the potential to distinguish "drug money" from "innocent" money. Because of the widely varying drug-use practices and patterns across countries and cultures, very different patterns of money contamination by drugs occur. Correlations of contamination with the source of money, however, have been weak. The degree of contamination is partly a function of the denomination of the banknote; in the USA, for example, denominations $5 through $50 have contained higher cocaine residue levels than $1 and $100 denominations. While banknote contamination can give an indication of types of drugs in use and especially recent proximity to bulk drug supplies, it has not provided insights on societal usage rates.

The forensics aspects of drug-contaminated money have been advanced largely by the work of investigators with Mass Spec Analytical Ltd. (MSA 2007). Overviews are available from Sleeman et al. (2000) and Armenta and de la Guardia (2008). Numerous papers have been published, a few of which are Bones et al. (2007b), Burton (1995), Carter et al. (2003), Ebejer et al. (2005, 2007), Jenkins (2001), Lavins et al. (2004), Luzardo et al. (2010), Sleeman et al. (1999), and Zuo et al. (2008).

This field will surely benefit from the rapid screening capabilities of ambient ionization mass spectrometry (e.g., Chen et al. 2009). Clearly, the potential exists for transfer of minute residues of illicit drugs from circulating money to the public; the ramifications of this, if any, are unknown.

6 Illicit Drugs in Ambient Air

Unlike the vast majority of pharmaceuticals, certain illicit drugs have the potential to escape to the ambient air, primarily because of the release of vapors and particulates from smoking and inhalation and from the generation of dusts; some of the only pharmaceuticals studied in air are the genotoxic chemotherapeutics used in the occupational setting (see references cited in Daughton and Ruhoy 2009). Perhaps the first data on an illicit drug in the environment were the 1998 report of cocaine associated with particulates in Los Angeles ambient outdoor air (Hannigan et al. 1998). Since then, studies have actively targeted a limited array of illicit drugs in ambient air in several locales, primarily cities in Italy and Spain, but also in Serbia, Portugal, Algeria, Chile, and Brazil.

An overview of this topic is provided by Postigo et al. (2010). The major studies include Balducci et al. (2009), Cecinato and Balducci (2007), Cecinato et al. (2009a, b, 2010), and Viana et al. (2010); another base of knowledge regarding analytical methodologies exists in the forensics literature, such as the work of Lai et al. (2008). Residues are usually associated with airborne particulates. Concentrations of cocaine generally are in the low picograms per cubic meter but can range up to low nanograms per cubic meter. Levels within a geographic region can vary by 2 or more orders of magnitude and are sensitive to weather conditions and time of year (with higher concentrations in winter) (Cecinato et al. 2010). These highest levels are roughly 3 orders of magnitude lower than commonly found for caffeine or nicotine. Also targeted in air studies have been other cocaine-related chemicals such as BZE and cocaethylene, as well as amphetamines, cannabinoids, cocainics, heroin, lysergics, methadone, and opioids. Multi-analyte air analysis has been rare, the work of Viana et al. (2010) being a recent example, with eight analytes targeted; this is one of the only reports of 6-AM in air.

The objective of air monitoring for illicit drugs is more in line with forensics (as a tool in detecting trends in drug usage) than with concerns regarding public health impacts from chronic pulmonary exposure to trace ambient levels. This is because cumulative lifetime doses (for example, with cocaine), even in locales with higher contamination, are 2–3 orders of magnitude below that of a single recreational dose (Cecinato et al. 2010; Viana et al. 2010). Atmospheric levels of illicit drugs, however, may be more transient and variable than levels in wastewater, adding greater complexity to its use as a tracking tool for drug usage.

7 Other Routes of Illicit Drug Impact on the Environment

7.1 Clan Labs

Clandestine drug laboratories (clan labs) are a primary localized source of certain drugs to the environment. Acute and chronic human health risks have been documented via all major exposure routes: inhalation, dermal absorption, and ingestion. Clan labs have been a recognized environmental hazard since the late 1980s

(Gardner 1989). Direct and collateral environmental impacts even from ephemeral production sites and facilities can be extensive (Cohen et al. 2007). Damage can result from negligent dumping of hazardous reagents and solvents, uncontrolled discharge of product chemicals and intermediates, alteration to watersheds (e.g., facilitation of erosion), and indiscriminate application of pesticides and fertilizers. In the USA, these impacts result primarily from production of cannabis and methamphetamine. Concerns are related not just to the synthesized parent drug (primarily methamphetamine in the USA) but also to the numerous synthesis starting materials and by-products (Snell 2001). With methamphetamine clan labs, a particularly problematic aspect is the insidious contamination of building structures (National Jewish Medical and Research Center 2005), in which large amounts of product permeate porous materials, creating reservoirs that serve as a perpetual source for future exposure. Morbidity from occupational and incidental human exposures is not trivial (Thrasher et al. 2009). The US EPA has issued new guidance for the cleanup of clan labs (USEPA 2009a).

Of particular interest is the financial liability and health risk posed by the purchase of contaminated real estate by unwary buyers (e.g., see Jarosz 2009; Poovey 2009). Methamphetamine-contaminated real estate has grown sufficiently common that it has fostered commercial enterprises specializing in the detection of methamphetamine (and other illicit drug) residues in real estate.

Worth noting is that wastewaters from pharmaceutical manufacturing facilities, which include both production and formulation facilities, had been largely ignored as a potential source of drug ingredients until the mid-2000s. The first survey of wastewaters from several manufacturing facilities in the USA revealed the presence of several drugs of abuse at levels over 1,000 μg/L (Phillips et al. 2010). Historically, reported levels of APIs have generally been 3 or more orders of magnitude lower than this in wastewater streams from municipalities not receiving manufacturing waste. This raises the possibility that in some locales pharmaceutical manufacturing could be a major source of certain drugs of abuse in ambient waters.

7.2 Livestock and Racing Animals

A wide spectrum of pharmaceuticals are known or suspected of being used illegally in livestock, primarily as growth promoters. An extensive literature exists on this subject, but due to the clandestine nature of the practice, an accurate picture does not exist for its full scope and magnitude, which probably varies greatly among countries. Some of these drugs are also abused by humans, so they can serve as another source contributing to environmental residue levels; others are unique to veterinary practice. Among the drugs in use, many may be registered for veterinary use but not for the purposes actually employed. Others may not be approved for any purpose. Included are members from the following classes: anthelmintics (e.g., levamisole), a wide range of antibiotics, coccidiostats (e.g., nitrofurans), hormones (anabolic steroids, corticosteroids, and thyreostats such as the thiouracils), β-agonists (e.g., clenbuterol), and tranquilizers (e.g., ketamine, haloperidol, xylazine) (Courtheyn et al. 2002; Stolker and Brinkman 2005).

Pharmaceuticals are known to contaminate much of the surroundings with which racehorses come into contact (or which their urine or sweat contacts), including stalls and racetracks (Barker 2008). Although the drugs detected in this monitoring study were primarily conventional non-steroidal anti-inflammatories (phenylbutazone, flunixin, and naproxen), analogous routes of contamination would not be unexpected for any illicit drug that may be surreptitiously used.

7.3 Dermal Contact and Transfer

Dermal transfer as a route of exposure for drugs has been an under-recognized aspect of drugs and the environment. The first comprehensive review of the ramifications of transfer of drugs from humans to the surfaces of any items contacted in the immediate surroundings (and to other people) by way of dermal transfer is provided by Daughton and Ruhoy (2009). There are two contributing factors. One is the transfer of residues remaining from topically applied drugs (which are generally applied at very high levels). The second is the excretion of systemic residues in sweat. Both factors apply equally to drugs of abuse and illicit drugs, especially potent analgesics such as fentanyl. The overall significance of this route of transfer to the immediate environment is not yet known.

7.4 Diversion

Diversion of licit drugs is the major route by which licit pharmaceuticals enter illicit markets and illicit use. Major routes include purchase from Internet pharmacies and theft from manufacturers, distributors, brick and mortar pharmacies, health-care facilities, and homes (e.g., for teen "pharming"). Pharmaceuticals still in clinical trials and not yet approved are even subject to diversion. A recent example is the selective androgen receptor modulator Andarine (a trifluoromethyl-arylpropionamide), which was being sold via the Internet to bodybuilders (Thevis et al. 2009).

Doctor/hospital shopping is also a form of diversion. A recent study of Internet pharmacies found that of nearly 3,000 online pharmacies (nearly half hosted in the USA), with combined annual sales of nearly US $12 billion, only 2 were certified by the Verified Internet Pharmacy Practice Sites (VIPPS) program, which is run by the National Association of Boards of Pharmacy (Felman 2009), and 10% stated that no prescription was required. Evidence points to diversion (as well as counterfeiting) as major sources for many of these drug stocks. The so-called rogue Internet pharmacies are documented as a significant source for diverted CPDs, especially Schedule III and Schedule IV drugs (NDIC 2009). Importation of drugs outside the regulatory system of the USA is a source of drugs with unknown magnitude. Estimates from the US Food and Drug Administration (FDA) have ranged from millions to tens of millions of packages of prescription drugs per year. These include counterfeit drugs, which include a wide array of undeclared active ingredients as well as undocumented designer drugs. Importation is a complex issue. An overview is provided by the US Government Accountability Office (USGAO 2005).

In addition to widespread outlets for illegally purchasing drugs of abuse, abusers have created a wide array of methods for "legally" diverting drugs. These include not just "doctor shopping" but also "hospital shopping." The latter is a practice in the USA that involves using free emergency services to acquire drugs to support addiction (Sullivan 2009).

7.5 Disposal of Leftover Medications

One particular aspect of drug occurrence in the environment can add significant confusion to assessing whether the source is from illicit or legal usage. For those drugs that share both legal and illicit usage (namely, those controlled substances not listed on DEA's Schedule I), a potentially major route by which their active ingredients can directly enter the environment is by flushing into sewers. While prudent practice for disposal of leftover drugs has generally shifted away from flushing (a practice long favored in order to reduce the incidence of intentional and unintended poisonings in the home), current guidance in the USA still recommends flushing a select list of drugs. As of June 2010, this list comprised 27 drugs, all of which are commonly abused or that pose inordinate risks of poisoning and therefore are hazardous if disposed into trash; they primarily contain the active ingredients fentanyl, hydromorphone, meperidine, methadone, morphine, and oxycodone (USFDA 2009). Some of these drugs (especially fentanyl) are formulated in delivery devices such as transdermal patches. After these devices have been expended, a significant portion of the active ingredient remains. These devices often contain large amounts of active ingredient. A used drug device can contribute quantities of the active ingredient that would exceed the amount that would otherwise be excreted after oral dosage. This is explained in Daughton and Ruhoy (2009).

8 Illicit Drugs and Environmental Impact

With the exception of the immediate and overt and hidden environmental impacts from clan labs, little is known about the potential actions of illicit drugs in the environment.

8.1 Fate and Transport

Compared with pharmaceuticals, little attention has been devoted to the environmental fate and transport of illicit drugs. Most illicit drugs have never been monitored in biosolids or sediments. Domènech et al. (2009) used fugacity modeling to predict the fate of cocaine and BZE. The microbial degradation of methamphetamine has been reported by Janusz et al. (2003). Wick et al. (2009) examined biological removal in activated sludge and found rapid removal for morphine, codeine, dihydrocodeine, oxycodone, and methadone but not for tramadol.

In two studies, the sorption of illicit drugs to sediments was reported (Stein et al. 2008; Wick et al. 2009). Wick et al. (2009) and Barron et al. (2009) acquired low distribution coefficients (Kd) for amphetamine, cocaine, cocaethylene, BZE, MDMA, morphine, codeine, dihydrocodeine, methadone, and tramadol, showing that removal via sorption to sewage sludge is possibly negligible.

8.2 Ecotoxicology

Far more is known regarding the ecotoxicology of licit pharmaceuticals than of illicit drugs, especially with regard to low-level mixed-stressor exposures. Almost nothing is known regarding the potential for biological effects in aquatic systems or the bioconcentration in biota of illicit drugs. Aquatic exposures are the primary focus.

To date, bioconcentration data for drugs of abuse have been reported in two studies. Diazepam is one of the only drugs with substantial illicit usage whose presence has been targeted in aquatic tissues. Diazepam was detected in all 10 fish liver samples analyzed from turbot at wet-weight concentrations ranging from 23 to 110 ng/g (Kwon et al. 2009). Diazepam is commonly detected in wastewaters from slaughterhouses (in China), albeit at low levels up to 16 ng/L (Shao et al. 2009),which shows that its illicit use extends beyond humans. Tramadol has been reported in the plasma of fish (up to 1.9 ng/g) exposed to treated sewage effluent (Fick et al. 2010).

The potential for effects from low-level exposure of fish is further complicated by the complexities in extrapolating across species. Data from the first in-depth study of an ectotherm with any analgesic (i.e., morphine) comport with extreme variability between species (Newby et al. 2006).

Gagne et al. (2006) report some nominal effects data from morphine in mussels. Scott et al. (2003) reported on the absence of adverse effects on soil microbial enzyme activity by six substances used in amphetamine synthesis, including P2P (phenyl-2-propanone), ephedrine, methamphetamine, and 3,4-methylenedioxybenzaldehyde.

Pharmacological studies of biological endpoints at ultra-low doses have relevance to the potential for both human and ecological effects from exposure to ambient residues in the environment, especially drinking water. Some of the pioneering studies relevant to ultra-low doses were conducted in the early 1990s and showed that biological effects could be obtained at doses many orders of magnitude lower than therapeutic doses; one example is the work of Crain and Shen (1995), who reported on the nociception in mice treated with doses as low as the femtomolar range. The subject of ultra-low dose effects has been discussed with respect to exposure to pharmaceuticals in drinking water (Daughton 2010 – in press).

9 The Future

Future work to address the various environmental aspects of illicit drugs in the environment would benefit from a comprehensive assessment of what has been accomplished to date and what new research is needed. Although the knowledge base regarding all aspects of illicit drugs in the environment is extremely small compared with that of pharmaceuticals, the body of published data is perhaps sufficiently large that we risk duplication of efforts while failing to address the more important remaining gaps or needs (Daughton 2009a). The first step in ensuring better-targeted research could be the creation of a centralized, publicly accessible database of results from research conducted worldwide. Such data should include both environmental occurrence data and data of absence (covering compartments such as sewage influent and effluent, sludge/biosolids, surface water, groundwater, and drinking water, air, wildlife tissues, and money), ecotoxicology (both field and controlled exposures), and especially data generated from FEUDS studies; metadata such as GIS (geographic information system), sampling and analytical methodologies, quality assurance, detection limits, and measures of range or variance are essential.

9.1 Advancing the Utility of FEUDS

Advancement of FEUDS as a topic of research as well as a population-level survey tool could occur on two fronts. First, numerous improvements could be made to better define and control the many variables contributing to uncertainty in FEUDS back-calculations for gauging collective drug usage. Standardized methodologies are needed, with better understood and controlled sources of error. The methodologies currently used for analysis of environmental samples for illicit drug ingredients span a wide range; this can be readily seen just for amphetamine and methamphetamine (e.g., see Boles and Wells 2010). Standardized methods are especially important for facilitating more meaningful inter-comparison of FEUDS data. Data from FEUDS studies also need to be assessed more rigorously against more comprehensive user surveys to better understand the accuracy and value of both approaches.

For FEUDS to succeed as a tool in gauging illicit drug usage for epidemiologic or forensic purposes, one variable in particular needs to be better understood – the pharmacokinetics (PK) of each drug, especially as it pertains to the excretion of unchanged parent drug and metabolites (especially conjugates); the importance of thoroughly understanding PK and conjugate excretion has been addressed by Daughton and Ruhoy (2009). PK parameters are key to accurate dose reconstruction. Although excretion rates for many pharmaceuticals are not well defined, even less is known about the PK of illicit drugs. PK and its poorly defined variability within a population contribute great uncertainty to the back-calculations used with FEUDS. Many factors contribute to the broad range of expression in population PK; genetic variability (such as single nucleotide polymorphisms) may lead to inter-occasion variability for the individual – partly as a function of environmental influences and physiological rhythms. The role of pharmacokinetics and environmental influence on drug metabolism is discussed in Daughton and Ruhoy (2009, 2010).

A comprehensive sensitivity analysis (which has yet to be performed) could possibly reveal that small changes in variables such as excretion rates (especially for extensively metabolized drugs) can lead to large errors in FEUDS calculations. For those drugs/metabolites with highly variable excretion rates, the error range could be substantial. As a case in point, with a study of 12 methamphetamine addicts, the urine ratio of amphetamine/methamphetamine ranged over 2 orders of magnitude – from 0.03 to 0.56 (Kim et al. 2008). This would also prove problematic for allocating amphetamine loadings in sewage to methamphetamine use versus medical use. A host of factors contribute to PK variability, including route and size of dose, gender, age, body mass, kidney and liver function, chronobiology, diet, polypharmacy interactions, and genetics/epigenetics (namely pharmacogenomics, which dictates the spectrum of PK variability). Similarly, it is important to be able to distinguish bacterial transformations in sewage (and the ambient environment) from those of human metabolism (Boleda et al. 2009).

Other potential ways to reduce errors in FEUDS calculations could be viewed as analogous to using internal correction methods such as internal standardization and isotope dilution. For example, instead of using correction factors based on modeling assumptions for dilution by waste streams and sewage transformations, correction factors could possibly be empirically derived by monitoring for particular pharmaceuticals. Pharmaceuticals that would be most useful for "calibrating" a WWTP system would be those that (i) are widely prescribed, (ii) are not abused or used recreationally, (iii) have real-time prescription sales data, (iv) are known to have high patient compliance (minimal leftovers, resulting in little disposal into sewers) and are used in short-term courses (not maintenance medications), (v) have a profile similar to that of the target illicit drug with regard to biodegradation and sorption to sewage solids, and (vi) have well-understood pharmacokinetics (preferably poorly metabolized, resulting in extensive excretion unchanged). By comparing the known consumption rates of the pharmaceutical "calibrant" (from prescribing databases) with the levels actually detected in the sewage stream, more accurate correction factors could possibly be derived and then applied to the illicit drug. By gathering long-term time-course data for the calibrant pharmaceutical, additional uncertainty could possibly be removed from the calibration factor. An example of a substance that may prove useful as a calibrant could be a metabolically refractory pharmaceutical such as iopromide – a widely used x-ray contrast agent with ubiquitous presence in sewage and natural waters. This approach, however, cannot remove the confounding of dual inputs from excretion and disposal of the targeted illicit drug; the latter, however, probably leads to episodic spikes in underlying baseline levels, which would become clearer with sustained monitoring.

The second front for improving the utility of FEUDS would be to expand its scope to tackle questions other than simply monitoring or gauging illicit drug consumption. Unexplored possibilities range from early detection of emerging trends in abuse of mainstream pharmaceuticals and in their illegal trafficking (e.g., from diversion or Internet purchases) to better gauging medication compliance rates for patients. For example, with access to real-time, local prescription data, those pharmaceutical ingredients in sewage whose back-calculated usage rates are substantially higher than the prescribed rates could be targeted for investigating the possibility of illegal trafficking. A possible example can be seen in the data presented by Kasprzyk-Hordern et al. (2009b; see Table 7 therein), in which calculated usage rates for more than two dozen prescribed and OTC pharmaceuticals are compared with known nationwide (not local) dispensing rates. Of these drugs, the calculated average usage rates exceeded the national average sales by over an order of magnitude for only one drug – tramadol. Indeed, tramadol (an opioid) is recognized for its growing incidence of misuse and abuse. Real-time prescription data are greatly confounded, however, by the inability of current tracking systems to correlate location of dispensing with place of actual use (e.g., because of transient populations and mail-order prescribing) (Ekedahl and Lindberg 2005). Another expanding source of data that could potentially be used to ground truth calculated usage rates is the growing network of collection programs that take back leftover consumer medications (see Glassmeyer et al. 2009).

An important aspect of FEUDS is that it has set the foundation for the use of sewage monitoring for other purposes – some unrelated to drug use. A fascinating possibility would be the use of sewage monitoring for measuring indicators of community-wide health status via the presence of various biomarkers of health or disease (discussed below).

9.2 Real-Time Monitoring of Community-Wide Health and Disease: Using Sewage Information Mining (SIM)

Within sewage is hidden a wealth of highly complex but chaotic chemical information about myriad aspects of biological processes. In the last 5 years, we have witnessed probably only the beginning of the applications for which sewage data could prove useful, namely FEUDS. Possibly first noted in 2008, Zuccato et al. (2008a) briefly mentioned that monitoring sewage "has the potential to extract useful epidemiologic data from qualitative and quantitative profiling of biological indicators entering the sewage system."

Perhaps the most important information contained in sewage resides with the countless biomarkers – substances that could serve as collective measures of community-wide health or disease. Biomarkers could serve as composite measures of exposure, stress, vulnerability to disease or overt disease, or health. Biomarkers include endogenous biochemicals produced in response to stress or indicative of health; they also include adducts of endogenous chemicals and xenobiotics. And of course, they include metabolites of significant detoxication or intoxication processes from xenobiotic exposure. Suitable markers could not have pharmaceutical equivalents, which would add great complexity to the modeling process because of the need to distinguish natural from anthropogenic sources; an example of an endogenous biomarker that has exogenous pharmacological use is cortisol (hydrocortisone).

As community-wide measures of health or disease status, a new discipline of SIM could provide, for the first time, the ability to gauge collective population-wide health and disease in real time. SIM would constitute the first true application of sewage chemistry to epidemiology and provide a means for conducting epidemiology in near-real time. SIM could also create the opportunity to view communities from a new perspective – "communities as the patient" – perhaps eventually leading to the paradigm of combining human and ecological communities as a single patient – as an interconnected whole. SIM could greatly expand our limited abilities for examining associations between human health and a host of environmental variables and stressors. It could hold the potential for greatly reducing the time and expense involved with establishing linkages between human disease and any stress imposed by the environment – or for gauging the effectiveness of new health-care measures. SIM could prove invaluable in more efficiently informing and targeting limited health-care resources. Illicit drugs have certainly provided insights for new ways to monitor the health of entire populations.

10 Summary

The published literature that addresses the many facets of pharmaceutical ingredients as environmental contaminants has grown exponentially since the 1990s. Although there are several thousand active ingredients used in medical pharmaceuticals worldwide, illicit drug ingredients (IDIs) have generally been excluded from consideration. Medicinal and illicit drugs have been treated separately in environmental research even though they pose many of the same concerns regarding the potential for both human and ecological exposure. The overview presented here covers the state of knowledge up until mid-2010 regarding the origin, occurrence, fate, and potential for biological effects of IDIs in the environment.

Similarities exist with medical pharmaceuticals, particularly with regard to the basic processes by which these ingredients enter the environment – excretion of unmetabolized residues (including via sweat), bathing, disposal, and manufacturing. The features of illicit drugs that distinguish them from medical pharmaceuticals are discussed. Demarcations between the two are not always clear, and a certain degree of overlap adds additional confusion as to what exactly defines an illicit drug; indeed, medical pharmaceuticals diverted from the legal market or used for non-medicinal purposes are also captured in discussions of illicit drugs. Also needing consideration as part of the universe of IDIs are the numerous adulterants and synthesis impurities often encountered in these very impure preparations. Many of these extraneous chemicals have high biological activity themselves.

In contrast to medical pharmaceuticals, comparatively little is known about the fate and effects of IDIs in the environment. Environmental surveys for IDIs have revealed their presence in sewage wastewaters, raw sewage sludge and processed sludge (biosolids), and drinking water. Nearly nothing is known, however, regarding wildlife exposure to IDIs, especially aquatic exposure such as indicated by bioconcentration in tissues. In contrast to pharmaceuticals, chemical monitoring surveys have revealed the presence of certain IDIs in air and monetary currencies – the latter being of interest for the forensic tracking of money used in drug trafficking. Another unknown with regard to IDIs is the accuracy of current knowledge regarding the complete scope of chemical identities of the numerous types of IDIs in actual use (particularly some of the continually evolving designer drugs new to forensic chemistry) as well as the total quantities being trafficked, consumed, or disposed.

The major aspect unique to the study of IDIs in the environment is making use of their presence in the environment as a tool to obtain better estimates of the collective usage of illicit drugs across entire communities. First proposed in 2001, but under investigation with field applications only since 2005, this new modeling approach for estimating drug usage by monitoring the concentrations of IDIs (or certain unique metabolites) in untreated sewage has potential as an additional source of data to augment or corroborate the information-collection ability of conventional written and oral surveys of drug-user populations. This still evolving monitoring tool has been called "sewer epidemiology" but is referred to in this chapter by a more descriptive proposed term "FEUDS" (Forensic Epidemiology Using Drugs in Sewage). The major limitation of FEUDS surrounds the variables involved at various steps performed in FEUDS calculations. These variables are summarized and span sampling and chemical analysis to the final numeric calculations, which particularly require a better understanding of IDI pharmacokinetics than currently exists. Although little examined in the literature, the potential for abuse of FEUDS as a tool in law enforcement is briefly discussed.

Finally, the growing interest in FEUDS as a methodological approach for estimating collective public usage of illicit drugs points to the feasibility of mining other types of chemical information from sewage. On the horizon is the potential for "sewage information mining" (SIM) as a general approach for measuring a nearly limitless array of biochemical markers that could serve as collective indicators of the specific or general status of public health or disease at the community-wide level. SIM may create the opportunity to view communities from a new perspective – "communities as the patient." This could potentially lead to the paradigm of combining human and ecological communities as a single patient – as an interconnected whole.

U.S. EPA Notice: The United States Environmental Protection Agency through its Office of Research and Development funded and managed the research described here. It has been subjected to Agency's administrative review and approved for publication.

References

Aaron R, Lewis P (1987) Cocaine residues on money. Crime Lab Dig 14: 18.

AMA (2009) Use of cannabis for medicinal purposes. Council on Science and Public Health (CSAPH), American Medical Association, CSAPH Report 3, http://americansforsafeaccess.org/downloads/AMA_Report.pdf.

Armenta S, de la Guardia M (2008) Analytical methods to determine cocaine contamination of banknotes from around the world. Trends Anal Chem 27: 344–351.

ATSDR (2009) Exposure-dose reconstruction program (EDRP). Web Page maintained by agency for toxic substances and disease registry (ATSDR), DHS, Atlanta, GA. http://www.atsdr.cdc.gov/edrp/.

Balducci C, Nervegna G, Cecinato A (2009) Evaluation of principal cannabinoids in airborne particulates. Anal Chim Acta 641: 89–94.

Banta-Green CJ, Field JA, Chiaia AC, Sudakin DL, Power L, de Montigny L (2009) The spatial epidemiology of cocaine, methamphetamine and 3,4-methylenedioxymethamphetamine (MDMA) use: a demonstration using a population measure of community drug load derived from municipal wastewater. Addiction 104: 1874–1880.

Barker SA (2008) Drug contamination of the equine racetrack environment: a preliminary examination. J Vet Pharmacol Ther 31: 466–471.

Barron L, Havel J, Purcell M, Szpak M, Kelleher B, Paull B (2009) Predicting sorption of pharmaceuticals and personal care products onto soil and digested sludge using artificial neural networks. Analyst 134: 663–670.

Bartelt-Hunt SL, Snow DD, Damon T, Shockley J, Hoagland K (2009) The occurrence of illicit and therapeutic pharmaceuticals in wastewater effluent and surface waters in Nebraska. Environ Pollut 157: 786–791.

Batt AL, Kostich MS, Lazorchak JM (2008) Analysis of ecologically relevant pharmaceuticals in wastewater and surface water using selective solid–phase extraction and UPLC-MS/MS. Anal Chem 80: 5021–5030.

BfR (2009) No health risk from the cocaine content in Red Bull Simply Cola [Kein Gesundheitsrisiko durch den Cocaingehalt in Red Bull Simply Cola]. Federal Institute for Risk Assessment (BfR: Bundesinstitut für Risikobewertung), BfR Health Assessment No. 020/2009, Berlin, Germany, 7 pp; http://www.bfr.bund.de/cm/245/no_health_risk_from_the_cocaine_content_in_red_bull_simply_cola.pdf.

Bijlsma L, Sancho JV, Pitarch E, Ibáñez M, Hernández F (2009) Simultaneous ultra-high-pressure liquid chromatography-tandem mass spectrometry determination of amphetamine and amphetamine-like stimulants, cocaine and its metabolites, and a cannabis metabolite in surface water and urban wastewater. J Chromatogr 1216: 3078–3089.

Bohannon J (2007) Hard data on hard drugs, grabbed from the environment: fieldwork in new and fast-growing areas of epidemiology requires wads of cash and a familiarity with sewer lines. Science 316: 42–44.

Boleda MR, Galceran MT, Ventura F (2007) Trace determination of cannabinoids and opiates in wastewater and surface waters by ultra-performance liquid chromatography-tandem mass spectrometry. J Chromatogr 1175: 38–48.

Boleda MR, Galceran MT, Ventura F (2009) Monitoring of opiates, cannabinoids and their metabolites in wastewater, surface water and finished water in Catalonia, Spain. Water Res 43: 1126–1136.

Boles TH, Wells MJM (2010) Analysis of amphetamine and methamphetamine as emerging pollutants in wastewater and wastewater-impacted streams. J Chromatogr 1217: 2561–2568.

Bones J, Thomas KV, Paull B (2007a) Using environmental analytical data to estimate levels of community consumption of illicit drugs and abused pharmaceuticals. J Environ Monit 9: 701–707.

Bones J, Macka M, Paull B (2007b) Evaluation of monolithic and sub 2 μm particle packed columns for the rapid screening for illicit drugs – application to the determination of drug contamination on Irish euro banknotes. Analyst 132: 208–217.

Boxall AB, Johnson P, Smith EJ, Sinclair CJ, Stutt E, Levy LS (2006a) Uptake of veterinary medicines from soils into plants. J Agric Food Chem 54: 2288–2297.

Boxall ABA et al. (2006b) Targeted monitoring study for veterinary medicines in the environment. Environment Agency, SC030183/SR, Bristol, England, http://publications. environmentagency.gov.uk/pdf/SCHO0806BLHH-e-e.pdf.

Burton F (1995) A study of the background levels of a range of controlled substances on Sterling banknotes in general circulation in England and Wales, Masters Dissertation, University of Bristol, Bristol, 121 pp.

Cai R, Crane E, Poneleit K, Paulozzi L (2010) Emergency department visits involving nonmedical use of selected prescription drugs—United States, 2004–2008. Morb Mortal Weekly Rep 59: 705–709.

Calisto V, Esteves VI (2009) Psychiatric pharmaceuticals in the environment. Chemosphere 77: 1257–1274.

Carter JF, Sleeman R, Parry J (2003) The distribution of controlled drugs on banknotes via counting machines. Forensic Sci Int 132: 106–112.

Castiglioni S, Zuccato E, Crisci E, Chiabrando C, Fanelli R, Bagnati R (2006) Identification and measurement of illicit drugs and their metabolites in urban wastewater by liquid chromatography – tandem mass spectrometry. Anal Chem 78: 8421–8429.

Castiglioni S, Zuccato E, Chiabrando C, Fanelli R, Bagnati R (2007) Detecting illicit drugs and metabolites in wastewater using high performance liquid chromatography-tandem mass spectrometry. Spectrosc Eur 19: 7–9.

Castiglioni S, Zuccato E, Chiabrando C, Fanelli R, Bagnati R (2008) Mass spectrometric analysis of illicit drugs in wastewater and surface water. Mass Spectrom Rev 27: 378–394.

Cecinato A, Balducci C (2007) Detection of cocaine in the airborne particles of the Italian cities Rome and Taranto. J Sep Sci 30: 1930–1935.

Cecinato A, Balducci C, Nervegna G (2009a) Occurrence of cocaine in the air of the World's cities: An emerging problem? A new tool to investigate the social incidence of drugs? Sci Total Environ 407: 1683–1690.

Cecinato A, Balducci C, Nervegna G, Tagliacozzo G, Allegrini I (2009b) Ambient air quality and drug aftermaths of the Notte Bianca (White Night) holidays in Rome. J Environ Monit 11: 200–204.

Cecinato A, Balducci C, Budetta V, Pasini A (2010) Illicit psychotropic substance contents in the air of Italy. Atmos Environ 44: 2358–2363.

Chapman S (2009) Consolidated Index of Drugs and Substances. Web Page maintained by Isomer Design, Toronto, Ontario. http://www.isomerdesign.com/Cdsa/scheduleNDX.php.

Chen H, Gamez G, Zenobi R (2009) What can we learn from ambient ionization techniques? J Am Soc Mass Spectrom 20: 1947–1963.

Chiaia AC, Banta-Green C, Field J (2008) Eliminating solid phase extraction with large-volume injection LC/MS/MS: Analysis of illicit and legal drugs and human urine indicators in US wastewaters. Environ Sci Technol 42: 8841–8848.

Chiuminatto U et al. (2010) Automated online solid phase extraction ultra high performance liquid chromatography method coupled with tandem mass spectrometry for determination of forty-two therapeutic drugs and drugs of abuse in human urine. Anal Chem 82: 5636–5645.

Christensen AM, Markussen B, Baun A, Halling-Sørensen B (2009) Probabilistic environmental risk characterization of pharmaceuticals in sewage treatment plant discharges. Chemosphere 77: 351–358.

Coetsier CM, Spinelli S, Lin L, Roig B, Touraud E (2009) Discharge of pharmaceutical products (PPs) through a conventional biological sewage treatment plant: MECs vs PECs? Environ Int 35: 787–792.

Cohen K, Sanyal N, Reed G (2007) Methamphetamine production on public lands: Threats and responses. Soc Nat Resour 20: 261–270.

Courtheyn D et al. (2002) Recent developments in the use and abuse of growth promoters. Anal Chim Acta 473: 71–82.

Crain SM, Shen K-F (1995) Ultra-low concentrations of naloxone selectively antagonize excitatory effects of morphine on sensory neurons, thereby increasing its antinociceptive potency and attenuating tolerance/dependence during chronic cotreatment. Proc Natl Acad Sci USA 92: 10540–10544.

Cunningham RT et al. (2009) Feasibility of a clinical chemical analysis approach to predict misuse of growth promoting hormones in cattle. Anal Chem 81: 977–983.

Daughton CG (2001a) Pharmaceuticals and personal care products in the environment: Overarching issues and overview. In: Daughton CG, Jones-Lepp TL (eds) Pharmaceuticals and personal care products in the environment: Scientific and regulatory issues. ACS Symposium Series 791, American Chemical Society, Washington, DC, Chapter 1, pp 2–38; doi:10.1021/bk-2001-0791.ch001; http://www.epa.gov/nerlesd1/bios/daughton/book-summary.htm.

Daughton CG (2001b) Literature forensics? Door to what was known but now forgotten. Environ Forensics 2: 277–282.

Daughton CG (2001c) Illicit drugs in municipal sewage: Proposed new non-intrusive tool to heighten public awareness of societal use of illicit/abused drugs and their potential for ecological consequence. In: Daughton CG, Jones-Lepp T (eds) Pharmaceuticals and personal care products in the environment: Scientific and regulatory issues. ACS Symposium Series 791, American Chemical Society, Washington, DC, Chapter 20, pp 348–364; doi:10.1021/bk-2001-0791.ch020; http://www.epa.gov/nerlesd1/bios/daughton/book-conclude.htm.

Daughton CG (2001d) Commentary on illicit drugs in the environment: a tool for public education – societal drug abuse and its aiding of terrorism. USEPA, NERL, Las Vegas, NV, 23 October, http://www.epa.gov/nerlesd1/bios/daughton/book-post.htm.

Daughton CG (2004) Non-regulated water contaminants: emerging research. Environ Impact Assess Rev 24: 711–732.

Daughton CG (2009a) Chemicals from the practice of healthcare: challenges and unknowns posed by residues in the environment. Environ Toxicol Chem 28: 2490–2494.

Daughton CG (2009b) Peering into the shadows of chemical space. Emerging contaminants and environmental science: is either being served by the other? Paper presented at 2nd International Conference on Occurrence, Fate, Effects, and Analysis of Emerging Contaminants in the Environment (EmCon09), opening address 4–7 August 2009, Fort Collins, CO; http://www.epa.gov/esd/bios/daughton/Daughton-abstract-EmCon09.pdf.

Daughton CG (2010) Pharmaceutical ingredients in drinking water: overview of occurrence and significance of human exposure. In: Halden R (ed) Emerging contaminants: Pharmaceuticals, personal care products. ACS Symposium Series 791. American Chemical Society, Washington, DC; see: http://pubs.acs.org/page/books/symposiumSeries/2010titles.html

Daughton CG (2011) Illicit drugs and the environment. In: Castiglioni S, Zuccato E (eds) Mass spectrometric analysis of illicit drugs in the environment, Wiley; ISBN 978-0-470-52954-6. Chapter1; see: http://www.wiley.com/WileyCDA/WileyTitle/productCd-0470529547.html

Daughton CG, Ruhoy IS (2008) The afterlife of drugs and the role of pharmEcovigilance. Drug Saf 31: 1069–1082.

Daughton CG, Ruhoy IS (2009) Environmental footprint of pharmaceuticals – the significance of factors beyond direct excretion to sewers. Environ Toxicol Chem 28: 2495–2521.

Daughton CG, Ruhoy IS (2010) Reducing the ecological footprint of pharmaceutical usage: linkages between healthcare practices and the environment. In: Kümmerer K, Hempel M (eds) Green and sustainable pharmacy. Springer, Berlin Heidelberg, Germany, Chapter 6, pp 77–103; doi:10.1007/978-3-642-05199-9_6; http://www.springer.com/environment/environmental+management/book/978-3-642-05198-2.

Daughton CG, Ternes TA (1999) Pharmaceuticals and personal care products in the environment: Agents of subtle change? Environ Health Perspect 107: 907–938.

Domènech X, Peral J, Muñoz I (2009) Predicted environmental concentrations of cocaine and benzoylecgonine in a model environmental system. Water Res 43: 5236–5242.

Ebejer KA, Brereton RG, Carter JF, Ollerton SL, Sleeman R (2005) Rapid comparison of diacetyl-morphine on banknotes by tandem mass spectrometry. Rapid Commun Mass Spectrom 19: 2137–2143.

Ebejer KA, Lloyd GR, Brereton RG, Carter JF, Sleeman R (2007) Factors influencing the contamination of UK banknotes with drugs of abuse. Forensic Sci Int 171: 165–170.

Ekedahl A, Lindberg G (2005) Differences between drug utilisation estimates based on pharmacy sales and on purchases by the resident population. Pharm World Sci 27: 469–471.

EMCDDA (2007) In aquae veritas? First European meeting on drugs and their metabolites in waste water. European Monitoring Centre for Drugs and Drug Addiction, Lisbon, Portugal, April–June, http://www.emcdda.europa.eu/html.cfm/index31432EN.html.

EMCDDA (2009a) Illicit consumption of drugs and the law – Situation in the EU Member States. Web Page maintained by European monitoring centre for drugs and drug addiction (EMCDDA), Lisbon, Portugal. http://eldd.emcdda.europa.eu/html.cfm/index5748EN.html.

EMCDDA (2009b) Classification of controlled drugs. Web Page maintained by European Monitoring Centre for Drugs and Drug Addiction (EMCDDA), Lisbon, Portugal. http://eldd.emcdda.europa.eu/html.cfm/index5622EN.html.

EMCDDA (2010) Action on new drugs. Web Page maintained by European monitoring centre for drugs and drug addiction, Lisbon, Portugal. http://www.emcdda.europa.eu/drug-situation/new-drugs.

Everts S (2010) Fake pharmaceuticals: Those fighting against counterfeit medicines face increasingly sophisticated adversaries. Chem Eng News 88: 27–29.

Felix JR, Hammer R, Gardner EA (2008) Cocaine contamination of currency in Birmingham AL. Inquiro (University of Alabama, Birmingham) 2: 50–62.

Felman F (2009) Big pharma facing brandjacking battle: Study confirms that sales of questionable drugs in illicit online pharmacies and B2B exchange sites continue to rise, putting supply chains and consumers at risk. Pharmaceutical Processing January: 2–4; http://www.markmonitor.com/download/eprint/PharmaceuticalProcessing-Jan09.pdf.

Fick J, Lindberg RH, Parkkonen J, Arvidsson B, Tysklind M, Larsson DGJ (2010) Therapeutic levels of levonorgestrel detected in blood plasma of fish: results from screening rainbow trout exposed to treated sewage effluents. Environ Sci Technol 44: 2661–2666.

Freye E (2009) History of designer drugs. In: Freye E, Levy JV (eds) Pharmacology and abuse of cocaine, amphetamines, ecstasy and related designer drugs: A comprehensive review on their mode of action, treatment of abuse and intoxication. Springer, The Netherlands, Chapter 16, pp 183–189; doi:http://dx.doi.org/10.1007/978-90-481-2448-0_27.

Frost N, Griffiths P (2008) Assessing illicit drugs in wastewater: potential and limitations of a new monitoring approach, Insights Series No. 9, European monitoring centre for drugs and drug addiction (EMCDDA), Lisbon, Portugal 100 pp.

Gagne F, Blaise C, Fournier M, Hansen PD (2006) Effects of selected pharmaceutical products on phagocytic activity in Elliptio complanata mussels. Comp Biochem Physiol C Toxicol Pharmacol 143: 179–186.

Gardner G (1989) Illegal drug laboratories: a growing health and toxic waste problem. Pace Environmental Law Review, Pace University School of Law, 7, 193–212 pp; http://digitalcommons.pace.edu/cgi/viewcontent.cgi?article=1132&context=envlaw.

Ghasemi J, Niazi A (2005) Two- and three-way chemometrics methods applied for spectrophotometric determination of lorazepam in pharmaceutical formulations and biological fluids. Anal Chim Acta 533: 169–177.

Gheorghe A et al. (2008) Analysis of cocaine and its principal metabolites in waste and surface water using solid-phase extraction and liquid chromatography-ion trap tandem mass spectrometry. Anal Bioanal Chem 391: 1309–1319.

Glassmeyer ST et al. (2009) Disposal practices for unwanted residential medications in the United States. Environ Int 35: 566–572.

González-Mariño I, Quintana JB, Rodríguez I, Rodil R, González-Peñas J, Cela R (2009) Comparison of molecularly imprinted, mixed-mode and hydrophilic balance sorbents performance in the solid-phase extraction of amphetamine drugs from wastewater samples for liquid chromatography-tandem mass spectrometry determination. J Chromatogr 1216: 8435–8441.

González-Mariño I, Quintana JB, Rodríguez I, Cela R (2010) Determination of drugs of abuse in water by solid-phase extraction, derivatisation and gas chromatography-ion trap-tandem mass spectrometry. J Chromatogr 1217: 1748–1760.

Goodman RA, Munson JW, Dammers K, Lazzarini Z, Barkley JP (2003) Forensic epidemiology: law at the intersection of public health and criminal investigations. J Law Med Ethics 31: 684–700.

Greenwald G (2009) Drug decriminalization in portugal: lessons for creating fair and successful drug policies. Cato Institute, Washington, DC, http://www.cato.org/pubs/wtpapers/greenwald_whitepaper.pdf.

Gros M, Petrović M, Barceló D (2009) Tracing pharmaceutical residues of different therapeutic classes in environmental waters by using liquid chromatography/quadrupole-linear ion trap mass spectrometry and automated library searching. Anal Chem 81: 898–912.

Gros M, Petrović M, Ginebreda A, Barceló D (2010) Removal of pharmaceuticals during wastewater treatment and environmental risk assessment using hazard indexes. Environ Int 36: 15–26.

Hagerman E (2008) Your sewer on drugs. Popular Sci 272: 44–59.

Hannigan MP et al. (1998) Bioassay-directed chemical analysis of Los Angeles airborne particulate matter using a human cell mutagenicity assay. Environ Sci Technol 32: 3502–3514.

Hering CL (2009) Flushing the fourth amendment down the toilet: how community urinalysis threatens individuals privacy. Arizona Law Rev 53: 741–776; http://www.arizonalawreview.org/ALR2009/VOL2513/Hering.pdf.

Huerta-Fontela M, Galceran MT, Ventura F (2007) Ultraperformance liquid chromatography-tandem mass spectrometry analysis of stimulatory drugs of abuse in wastewater and surface waters. Anal Chem 79: 3821–3829.

Huerta-Fontela M, Galceran MT, Ventura F (2008a) Stimulatory drugs of abuse in surface waters and their removal in a conventional drinking water treatment plant. Environ Sci Technol 42: 6809–6816.

Huerta-Fontela M, Galceran MT, Martin-Alonso J, Ventura F (2008b) Occurrence of psychoactive stimulatory drugs in wastewaters in north-eastern Spain. Sci Total Environ 397: 31–40.

Huerta-Fontela M, Galceran MT, Ventura F (2010) Illicit drugs in the urban water cycle. In: Xenobiotics in the urban water cycle. Springer, Netherlands, Chapter 3, pp 51–71; doi:10.1007/978-90-481-3509-7_3; http://dx.doi.org/10.1007/978-90-481-3509-7_3.

Hummel D, Loffler D, Fink G, Ternes TA (2006) Simultaneous determination of psychoactive drugs and their metabolites in aqueous matrices by liquid chromatography mass spectrometry. Environ Sci Technol 40: 7321–7328.

INCB (2009) International narcotics control board: narcotic drugs, psychotropic substances, and precursors. Web Page maintained by International Narcotics Control Board, Vienna, Austria http://www.incb.org/.

Inoue H, Iwata YT, Kuwayama K (2008) Characterization and profiling of methamphetamine seizures. J Health Sci 54: 615–622.

Janusz A, Kirkbride KP, Scott TL, Naidu R, Perkins MV, Megharaj M (2003) Microbial degradation of illicit drugs, their precursors, and manufacturing by-products: Implications for clandestine drug laboratory investigation and environmental assessment. Forensic Sci Int 134: 62–71.

Jarosz F (2009) Scores of Indiana homes contaminated by meth labs sit abandoned: contaminated by meth production, scores of Indiana homes abandoned after labs are busted, no one enforces cleanup, Indystarcom, Community Star Network, Rome City, Ind., 10 May; http://www.indy.com/posts/scores-of-indiana-homes-contaminated-by-meth-labs-sit-abandoned.

Jenkins AJ (2001) Drug contamination of US paper currency. Forensic Sci Int 121: 189–193.

Jones-Lepp TL, Stevens R (2007) Pharmaceuticals and personal care products in biosolids/sewage sludge: The interface between analytical chemistry and regulation. Anal Bioanal Chem 387: 1173–1183.

Jones-Lepp TL, Alvarez DA, Petty JD, Huckins JN (2004) Polar organic chemical integrative sampling and liquid chromatography–electrospray/ion-trap mass spectrometry for assessing selected prescription and illicit drugs in treated sewage effluents. Arch Environ Contam Toxicol 47: 427–439.

Kaleta A, Ferdig M, Buchberger WS (2006) Semiquantitative determination of residues of amphetamine in sewage sludge samples. J Sep Sci 29: 1662–1666.

Karolak S, Nefau T, Bailly E, Solgadi A, Levi Y (2010) Estimation of illicit drugs consumption by wastewater analysis in Paris area (France). Forensic Sci Int 200: 153–160.

Kasprzyk-Hordern B, Dinsdale RM, Guwy AJ (2007) Multi-residue method for the determination of basic/neutral pharmaceuticals and illicit drugs in surface water by solid-phase extraction and ultra performance liquid chromatography-positive electrospray ionisation tandem mass spectrometry. J Chromatogr 1161: 132–145.

Kasprzyk-Hordern B, Dinsdale RM, Guwy AJ (2008a) Multiresidue methods for the analysis of pharmaceuticals, personal care products and illicit drugs in surface water and wastewater by solid-phase extraction and ultra performance liquid chromatography-electrospray tandem mass spectrometry. Anal Bioanal Chem 391: 1293–1308.

Kasprzyk-Hordern B, Dinsdale RM, Guwy AJ (2008b) The occurrence of pharmaceuticals, personal care products, endocrine disruptors and illicit drugs in surface water in South Wales, UK. Water Res 42: 3498–3518.

Kasprzyk-Hordern B, Dinsdale RM, Guwy AJ (2009a) The removal of pharmaceuticals, personal care products, endocrine disruptors and illicit drugs during wastewater treatment and its impact on the quality of receiving waters. Water Res 43: 363–380.

Kasprzyk-Hordern B, Dinsdale RM, Guwy AJ (2009b) Illicit drugs and pharmaceuticals in the environment – Forensic applications of environmental data, Part 1: Estimation of the usage of drugs in local communities. Environ Pollut 157: 1773–1777.

Kasprzyk-Hordern B, Kondakal VVR, Baker DR (2010) Enantiomeric analysis of drugs of abuse in wastewater by chiral liquid chromatography coupled with tandem mass spectrometry. J Chromatogr 1217: 4575–4586.

Khan SJ (2002) Occurrence, behavior and fate of pharmaceutical residues in sewage treatment, Doctoral Dissertation, University of New South Wales, New South Wales, Australia, 383 pp.

Khan S, Ongerth JE (2005) Occurrence and distribution of pharmaceutical residuals in bay sewage and sewage treatment (prepared for Bay Area Clean Water Agencies). University of New South Wales, School of Civil and Environmental Engineering, 8012–8017, 29 August, 75 pp; http://www.bacwa.org/LinkClick.aspx?fileticket=o%2F1vMUtfBeU%3D&tabid=101&mid=452.

Kim E et al. (2008) Comparison of methamphetamine concentrations in oral fluid, urine and hair of twelve drug abusers using solid-phase extraction and GC-MS. Annales de Toxicologie Analytique 20: 145–153.

Kortenkamp A, Backhaus T, Faust M (2009) State of the art report on mixture toxicity: Final report. University of London School of Pharmacy (ULSOP), 22 December, 391 pp; http://ec.europa.eu/environment/chemicals/pdf/report_Mixture%20toxicity.pdf.

Kwon J-W, Armbrust KL, Vidal-Dorsch D, Bay SM, Xia K (2009) Determination of 17α-ethynylestradiol, carbamazepine, diazepam, simvastatin, and oxybenzone in fish livers. J AOAC Int 92: 359–369.

Lai H, Corbin I, Almirall J (2008) Headspace sampling and detection of cocaine, MDMA, and marijuana via volatile markers in the presence of potential interferences by solid phase microextraction–ion mobility spectrometry (SPME-IMS). Anal Bioanal Chem 392: 105–113.

Lavins ES, Lavins BD, Jenkins AJ (2004) Cannabis (marijuana) contamination of United States and foreign paper currency. J Anal Toxicol 28: 439–442.

Loganathan B, Phillips M, Mowery H, Jones-Lepp TL (2009) Contamination profiles and mass loadings of select macrolide antibiotics and illicit drugs from a small urban wastewater treatment plant. Chemosphere 75: 70–77.

Loue S (2010) Forensic epidemiology: Integrating public health and law enforcement. Jones and Bartlett, Boston, MA 195 pp.

Luzardo OP, Zumbado M, Almeida-González M, Boada LD (2010) Evaluating habits of abuse of illicit drugs in a tourist region (the Canary Islands, Spain) through the determination of drug residues in Euro banknotes. Toxicol Lett 196: S290-S290.

Magura S (2010) Validating self-reports of illegal drug use to evaluate national drug control policy: A reanalysis and critique. Eval Program Plann 33: 234–237.

Mari F et al. (2009) Cocaine and heroin in waste water plants: A 1-year study in the city of Florence, Italy. Forensic Sci Int 189: 88–92.

McKellar QA (1997) Ecotoxicology and residues of anthelmintic compounds. Vet Parasitol 72: 413–435.

MSA (2007) Mass spec analytical Ltd: Papers Published [forensics of drug-contaminated money]. Web Page maintained by Mass Spec Analytical Ltd., Bristol; http://www.msaltd.co.uk/papers.htm.

National Jewish Medical and Research Center (2005) Chemical exposures associated with clandestine methamphetamine laboratories using the hypophosphorous and phosphorous flake method of production. National Jewish Medical and Research Center, Denver, CO, 23 September, 20 pp; http://www.nationaljewish.org/pdf/meth-hypo-cook.pdf.

NDIC (2009) Diversion of CPDs. National Prescription Drug Threat Assessment 2009, U.S. Department of Justice, National Drug Intelligence Center, 2010-Q0317-001, Johnstown, PA, April; http://www.justice.gov/ndic/pubs33/33775/diversion.htm; http://www.justice.gov/ndic/pubs33/33775/index.htm#Contents.

NDIC (2010) National drug threat assessment 2010. U.S. Department of Justice, National Drug Intelligence Center, 2010-Q0317-001, Johnstown, PA, February, http://www.justice.gov/ndic/pubs38/38661/index.htm.

Newby NC, Mendonça PC, Gamperl K, Stevens ED (2006) Pharmacokinetics of morphine in fish: winter flounder (Pseudopleuronectes americanus) and seawater-acclimated rainbow trout (Oncorhynchus mykiss). Comp Biochem Physiol C Toxicol Pharmacol 143: 275-283.

NIDA (2008) Monitoring the future survey. Web Page maintained by National Institute on Drug Abuse (NIDA), National Institutes of Health (NIH), http://www.nida.nih.gov/Drugpages/MTF.html.

NIDA (2009) Drugs of abuse information. Web Page maintained by National Institute on Drug Abuse (NIDA), National Institutes of Health (NIH), http://www.drugabuse.gov/drugpages/.

Nieto A, Peschka M, Borrull F, Pocurull E, Marcé RM, Knepper TP (2010) Phosphodiesterase type V inhibitors: Ocurrence[sic] and fate in wastewater and sewage sludge. Water Res 44: 1607–1615.

Noppe H, Le Bizec B, Verheyden K, De Brabander HF (2008) Novel analytical methods for the determination of steroid hormones in edible matrices. Anal Chim Acta 611: 1–16.

Nutt DJ (2009) Equasy – An overlooked addiction with implications for the current debate on drug harms. J Psychopharmacol 23: 3–5.

ONDCP (2009) Federal drug data sources. Web Page maintained by Office of National Drug Control Policy, http://www.whitehousedrugpolicy.gov/DrugFact/sources.html.

Oyler J, Darwin WD, Cone EJ (1996) Cocaine contamination of United States paper currency. J Anal Toxicol 20: 213–216.

Paulozzi LJ, Xi Y (2008) Recent changes in drug poisoning mortality in the United States by urban-rural status and by drug type. Pharmacoepidemiol Drug Saf 17: 997–1005.

Pedersen JA, Soliman M, Suffet IH (2005) Human pharmaceuticals, hormones, and personal care product ingredients in runoff from agricultural fields irrigated with treated wastewater. J Agric Food Chem 53: 1625–1632.

Phillips PJ et al. (2010) Pharmaceutical formulation facilities as sources of opioids and other pharmaceuticals to wastewater treatment plant effluents. Environ Sci Technol 44: 4910–4916.

Pichini S et al. (2008) Liquid chromatography-atmospheric pressure ionization electrospray mass spectrometry determination of "hallucinogenic designer drugs" in urine of consumers. J Pharm Biomed Anal 47: 335–342.

Poon WT, Lam YH, Lai CK, Chan AY, Mak TW (2007) Analogues of erectile dysfunction drugs: an under-recognised threat. Hong Kong Med J 13: 359–363.

Poovey B (2009) Meth makers leave behind a toxic trail at motels, Star Telegram, Associated Press, 23 February; http://abcnews.go.com/Business/wireStory?id=6937417.

Postigo C, Lopez de Alda MJ, Barceló D (2008a) Analysis of drugs of abuse and their human metabolites in water by LC-MS2. Trends Anal Chem 27: 1053–1069.

Postigo C, Lopez de Alda MJ, Barcelo D (2008b) Fully automated determination in the low nanogram per liter level of different classes of drugs of abuse in sewage water by on-line solid-phase extraction-liquid chromatography-electrospray-tandem mass spectrometry. Anal Chem 80: 3123–3134.

Postigo C et al. (2009) Determination of drugs of abuse in airborne particles by pressurized liquid extraction and liquid chromatography-electrospray-tandem mass spectrometry. Anal Chem 81: 4382–4388.

Postigo C, López de Alda MJ, Barceló D (2010) Drugs of abuse and their metabolites in the Ebro River basin: Occurrence in sewage and surface water, sewage treatment plants removal efficiency, and collective drug usage estimation. Environ Int 36: 75–84.

Psychonaut Web Mapping Research Group (2010) Psychonaut web mapping project: final report – alert on new recreational drugs on the web; building up a European-wide web scan-monitoring system. Institute of Psychiatry, King's College London, London, February, 17 pp; http://194.83.136.209/documents/reports/Psychonaut_Project_Executive_Summary.pdf also http://194.83.136.209/project.php.

Redshaw C, Cooke M, Talbot H, McGrath S, Rowland S (2008) Low biodegradability of fluoxetine HCl, diazepam and their human metabolites in sewage sludge-amended soil. J Soils Sed 8: 217–230.

Ruhoy IS, Daughton CG (2008) Beyond the medicine cabinet: An analysis of where and why medications accumulate. Environ Int 34: 1157–1169.

Scott TL, Janusz A, Perkins MV, Megharaj M, Naidu R, Kirkbride KP (2003) Effect of amphetamine precursors and by-products on soil enzymes of two urban soils. Bull Environ Contam Toxicol 70: 0824–0831.

Shao B, Chen D, Zhang J, Wu Y, Sun C (2009) Determination of 76 pharmaceutical drugs by liquid chromatography-tandem mass spectrometry in slaughterhouse wastewater. J Chromatogr 1216: 8312–8318.

Sleeman R, Burton IFA, Carter JF, Roberts DJ (1999) Rapid screening of banknotes for the presence of controlled substances by thermal desorption atmospheric pressure chemical ionisation tandem mass spectrometry. Analyst 124: 103–108.

Sleeman R, Burton F, Carter J, Roberts D, Hulmston P (2000) Drugs on money. Anal Chem 72: 397 A–403 A.

Smith SW (2009) Chiral toxicology: it's the same thing...only different. Toxicol Sci 110: 4–30.

Snell MB (2001) Welcome to meth country. Sierra 86: http://www.sierraclub.org/sierra/200101/Meth.asp.

Sommer C, Bibby BM (2002) The influence of veterinary medicines on the decomposition of dung organic matter in soil. Eur J Soil Biol 38: 155–159.

Sörgel F (2006) High cocaine use in Europe and US proven stunning data for European Countries: First ever comparative multi-country study of cocaine use by a new measurement technique. Institute for Biomedical and Pharmaceutical Research (IBMP), Nürnberg-Heroldsberg, Germany; http://www.sharedresponsibility.gov.co/en/download/drug_consumption/IMDB_Cocaine_River_Study_2006.pdf.

Stein K, Ramil M, Fink G, Sander M, Ternes TA (2008) Analysis and sorption of psychoactive drugs onto sediment. Environ Sci Technol 42: 6415–6423.

Stolker AAM, Brinkman UAT (2005) Analytical strategies for residue analysis of veterinary drugs and growth-promoting agents in food-producing animals – a review. J Chromatogr 1067: 15–53.

Straub JO (2008) Deterministic and probabilistic environmental risk assessment for diazepam. In: Kümmerer K (ed) Pharmaceuticals in the environment – sources, fate, effects and risks, 3rd ed. Springer, Berlin Heidelberg, Chapter 22, pp 343–383; doi:10.1007/978-3-540-74664-5_22; http://www.springerlink.com/content/r573826833631506/?p=9ed3dcab782d422fa25c2de4024 61f54&pi=21.

Sullivan J (2009) The ride to stay high: How drug addicts manipulate EMS, hospitals for their fix, The Ironton Tribune, Boone Newspapers, Inc., Ironton, OH; http://www.irontontribune.com/news/2009/jun/13/ride-stay-high/.

Sussman S, Huver RME (2006) Definitions of street drugs. In: Cole SM (ed) New research on street drugs. Nova Science Publishers, Inc., New York, NY, Chapter 1, pp 1–12.

Sussman S, Ames SL (2008) Concepts of drugs, drug use, misuse, and abuse. In: Sussman S, Ames SL (eds) Drug abuse: concepts, prevention, and cessation, Cambridge University Press, Cambridge, Chapter 1, pp 3–17; doi:10.1017.

Teijon G, Candela L, Tamoh K, Molina-Díaz A, Fernández-Alba AR (2010) Occurrence of emerging contaminants, priority substances (2008/105/CE) and heavy metals in treated wastewater and groundwater at Depurbaix facility (Barcelona, Spain). Sci Total Environ 408: 3584–3595.

Terzic S, Senta I, Ahel M (2010) Illicit drugs in wastewater of the city of Zagreb (Croatia) – Estimation of drug abuse in a transition country. Environ Pollut 158(8):2686–2693; doi 10.1016/j.envpol.2010.04.020:

Thevis M, Geyer H, Kamber M, Schänzer W (2009) Detection of the arylpropionamide-derived selective androgen receptor modulator (SARM) S-4 (Andarine) in a black-market product. Drug Test Anal 1: 387–392.

Thompson T (2002) £15m of notes tainted by drugs are destroyed, The Observer, Guardian News and Media Limited, UK, 10 November; http://www.guardian.co.uk/uk/2002/nov/10/drugsandalcohol.ukcrime.

Thrasher D, Von Derau K, Burgess J (2009) Health effects from reported exposure to methamphetamine labs: A poison center-based study. J Med Toxicol 5: 200–204.

UNODC (June 2007) 2007 World drug report: section 4 – methodology. United Nations Office on Drugs and Crime, Vienna, Austria, 272–274 pp; http://www.unodc.org/pdf/research/wdr07/WDR_2007_4.0_methodology.pdf.

UNODC (2009a) World drug report – global illicit drug trends. United Nations Office on Drugs and Crime, Vienna, Austria, http://www.unodc.org/unodc/data-and-analysis/WDR.html.

UNODC (2009b) Information about drugs. Web Page maintained by United Nations Office on Drugs and Crime, Vienna, Austria. http://www.unodc.org/unodc/en/illicit-drugs/definitions/index.html.

USDEA (2008) National Forensic Laboratory Information System (NFLIS) year 2008 annual report. National Forensic Laboratory System, US Drug Enforcement Administration, Office of Diversion Control, 32 pp; http://www.nflis.deadiversion.usdoj.gov/Reports/NFLIS2008AR.pdf.

USEPA (2009a) U.S. EPA voluntary guidelines for methamphetamine laboratory cleanup. In: US Environmental Protection Agency, Office of Solid Waste and Emergency Response, p 48.

USEPA (2009b) Pharmaceuticals and personal care products (PPCPs): relevant literature. Web Page maintained by US Environmental Protection Agency (a comprehensive database of literature references compiled by CG Daughton and MST Scuderi; first implemented 19 February 2008), Las Vegas, NV; http://www.epa.gov/ppcp/lit.html.

USFDA (2009) Disposal by flushing of certain unused medicines: What you should know. Web Page maintained by US Food and Drug Administration, Rockville, MD;

http://www.fda.gov/Drugs/ResourcesForYou/Consumers/BuyingUsingMedicineSafely/EnsuringSafeUseofMedicine/SafeDisposalofMedicines/ucm186187.htm.

USGAO (2005) Prescription drugs: Strategic framework would promote accountability and enhance efforts to enforce the prohibitions on personal importation. United States Government Accountability Office, Washington, DC, 8 September, 76 pp; http://www.gao.gov/new.items/d05372.pdf.

van Nuijs ALN et al. (2009a) Analysis of drugs of abuse in wastewater by hydrophilic interaction liquid chromatography-tandem mass spectrometry. Anal Bioanal Chem 395: 819–828.

van Nuijs ALN et al. (2009b) Cocaine and metabolites in waste and surface water across Belgium. Environ Pollut 157: 123–129.

van Nuijs ALN et al. (2009c) Can cocaine use be evaluated through analysis of wastewater? A nation-wide approach conducted in Belgium. Addiction 104: 734–741.

van Nuijs ALN et al. (2010 – in press) Illicit drug consumption estimations derived from wastewater analysis: A critical review. Sci Total Environ doi 10.1016/j.scitotenv.2010.05.030:

Vazquez-Roig P, Blasco C, Andreu V, Pascual JA, Rubio JL, Picó Y (2010) Water quality in coastal wetlands: Illicit drugs in surface waters of L'Albufera Natural Park (Valencia, Spain). Geophysical Research Abstracts 12: EGU2010-14490.

Venhuis BJ, de Kaste D (2008) Sildenafil analogs used for adulterating marihuana. Forensic Sci Int 182: e23–e24.

Venhuis BJ, Barends DM, Zwaagstra ME, de Kaste D (2007) Recent developments in counterfeits and imitations of Viagra, Cialis and Levitra: A 2005–2006 update. RIVM (Netherlands National Institute for Public Health and the Environment), RIVM Report 370030001/2007, Bilthoven, the Netherlands, 61 pp; http://rivm.openrepository.com/rivm/bitstream/10029/16459/1/370030001.pdf.

Verster JC (2010) Monitoring drugs of abuse in wastewater and air. Curr Drug Abuse Rev 3: 1–2.

Viana M et al. (2010) Drugs of abuse in airborne particulates in urban environments. Environ Int 36: 527–534.

WHO (2008) Counterfeit drugs kill. International Medical Products Anti-Counterfeiting Taskforce (IMPACT), World Health Organization, May, 8 pp; http://www.who.int/impact/FinalBrochureWHA2008a.pdf.

Wick A, Fink G, Joss A, Siegrist H, Ternes T (2009) Fate of beta blockers and psycho-active drugs in conventional wastewater treatment. Water Res 43: 1060–1074.

Wiergowski M, Szpiech B, Reguła K, Tyburska A (2009) Municipal sewage as a source of current information on psychoactive substances used in urban communities. Probl Forensic Sci 79: 327–337.

Yang Y, Shao B, Zhang J, Wu Y, Duan H (2009) Determination of the residues of 50 anabolic hormones in muscle, milk and liver by very-high-pressure liquid chromatography-electrospray ionization tandem mass spectrometry. J Chromatogr B 877: 489–496.

Zuccato E, Castiglioni S (2009) Illicit drugs in the environment. Philos Trans R Soc A Math Phys Eng Sci 367: 3965–3978.

Zuccato E et al. (2005) Cocaine in surface waters: New evidence-based tool to monitor community drug abuse. Environmental Health: A Global Access Science Source 4: 7 pp.

Zuccato E, Chiabrando C, Castiglioni S, Bagnati R, Fanelli R (2008a) Estimating community drug abuse by wastewater analysis. Environ Health Perspect 116: 1027–1032.

Zuccato E, Castiglioni S, Bagnati R, Chiabrando C, Grassi P, Fanelli R (2008b) Illicit drugs, a novel group of environmental contaminants. Water Res 42: 961–968.

Zuo Y, Zhang K, Wu J, Rego C, Fritz J (2008) An accurate and nondestructive GC method for determination of cocaine on US paper currency. J Sep Sci 31: 2444–2450.

Cumulative Subject Matter Index
Volumes 201–210

1,3-Butadiene, pulmonary toxin, **201**:52
10-Trifluoromethoxy-decane-1-sulfonate (TFM-DS), biodegradation pathways (diag.), **208**:172
2,4-Decadienal, pulmonary toxin, **201**:52
9-[4-(Trifluoromethyl)phenoxy]-nonane-1-sulfonate (TFMP-NS), biodegradation pathway (diag.), **208**:173

A

Abbreviations, used in this volume (table), **209**:4
Abiotic degradation, pharmaceuticals, **202**:127
Abiotic parameters, gammarid toxicity effects, **205**:18
Absorption & elimination, bioconcentration theory, **204**:19
Accidents worldwide, tanker oil spills (table), **206**:98
Accidents, Baltic Sea shipping, **206**:106
Accidents, oil spill causes (diag.), **206**:97
Accidents, shipping oil spills, **206**:95 ff.
Accumulation in earthworms, chemicals, **203**:35
Acetylacholinesterase (AChE) activity, neurotoxicity biomarker, **205**:50
Acetylcholinesterase inhibition, chlorpyrifos, **209**:11
Acetylcholinesterase inhibition, OP mode-of-action, **205**:119
AChE (acetylcholinesterase) inhibition, chlorpyrifos, **209**:11
Acquired immunity, aging effects, **207**:119
ACR (acute-to-chronic ratio) calculation, pesticide default values (table), **209**:69
Acrolein, pulmonary toxin, **201**:50
Acronyms, used in this volume (table), **209**:4
ACRs in criteria derivation, **209**:66

ACRs, chronic criteria derivation, **209**:114
ACRs, literature default values, **209**:116
ACRs, single-chemical multispecies, **209**:67
Acrylamide, dermatotoxicity, **203**:126
Active transport, chemicals, **203**:7
Acute aquatic criteria derivation, SSD procedure, **209**:106
Acute criterion derivation, AF procedure, **209**:113
Acute criterion derivation, UCDM, **209**:105
Acute data, to estimate chronic toxicity, **209**:16
Acute factor utilization, AF procedures, **209**:64
Acute poisoning, OP (organophosphorous) pesticides, **205**:132
Acute toxicity, fenamiphos in fish, (table), **205**:140
Acute toxicity, pesticide data sets (table), **209**:39
Acute-to-chronic ratios (ACRs), in criteria derivation, **209**:66
Acyltransferases, aquatic organisms, **204**:79
Additivity models, toxicity of mixtures, **209**:81
ADHD (attention deficit hyperactivity disorder), pollutants, **201**:2
Adulterants, in illicit drugs (table), **210**:69–70
Advanced glycation end products, α-dicarbonyl species, **204**:142
Adverse drug reactions, polypharmacy, **207**:132
Aerobic degradation pathway, polychlorinated biphenyls (PCBs) (diag.), **201**:143
Aerobic-activated sludge, N-EtFOSE fate (diag.), **208**:169
AF (assessment factor) procedure, criteria derivation, **209**:60
AF procedure, acute criterion derivation, **209**:113
AF procedures, acute factor utilization, **209**:64

D.M. Whitacre (ed.), *Reviews of Environmental Contamination and Toxicology*,
Reviews of Environmental Contamination and Toxicology 210,
DOI 10.1007/978-1-4419-7615-4, © Springer Science+Business Media, LLC 2011

AFs used, in existing methodologies (table), **209**:61
AFs, choosing toxicity values, **209**:63
AFs, proper use, **209**:62
AFs, setting magnitude factors, **209**:64
Aged residue effects, PCBs, **201**:139
Age-dependence, pesticide bioconcentration, **204**:14
Age-related changes, environmental chemical sensitivity, **207**:97
Aging effect, brain volume, **207**:110
Aging effects, acquired (adaptive) immunity, **207**:119
Aging effects, B-lymphocyte function, **207**:118
Aging effects, bone density decrease, **207**:108
Aging effects, cardiovascular system, **207**:102
Aging effects, cytokine function, **207**:118
Aging effects, endocrine system, **207**:122
Aging effects, environmental chemical skin penetration, **207**:129
Aging effects, hematopoiesis, **207**:106
Aging effects, hepatic cytochrome activity, **207**:102
Aging effects, integument & epidermis, **207**:126
Aging effects, menopausal changes in women (table), **207**:135
Aging effects, microsomal activity, **207**:101
Aging effects, neutrophil function, **207**:117
Aging effects, on hearing, **207**:113
Aging effects, on innate immunity, **207**:117
Aging effects, on macrophage function, **207**:117
Aging effects, on the sex hormone, **207**:125
Aging effects, parathyroid hormone, **207**:123
Aging effects, respiratory system, **207**:130
Aging effects, the immune system, **207**:116
Aging effects, the nervous system, **207**:109
Aging effects, urinary system changes, **207**:98
Aging effects, ventricular performance, **207**:105
Aging effects, vision (table), **207**:113
Aging effects, xenobiotic clearance, **207**:101
Aging effects, xenobiotic exposure, **207**:109
Aging humans, environmental chemical susceptibility, **207**:96
Aging, a definition, **207**:96
Aging, blood-brain barrier effects, **207**:111
Aging, glomerular-filtration effects, **207**:98
Aging, hepatic function changes, **207**:100
Aging, nerve effects, **207**:111
Aging, the kidney, **207**:97

Agricultural chemicals, Great Lakes pollution, **207**:14
AIDS patients, *Pseudomonas aeruginosa*, **201**:85
Air contamination, illicit drugs, **210**:91
Air criteria, criteria setting, **209**:92
Air criteria, harmonization with aquatic criteria, **209**:132
Air criteria, harmonization, **209**:92
Air residues, perfluorinated compounds, **208**:204
Alaska, lead-zinc mine, **206**:49
Alaskan lead exposure, adults & children, **206**:58
Alcohol, osteoporosis effects, **207**:109
Aldrin toxicity data, comparative distribution fit (illus.), **209**:45
Aldrin, data-set distribution test (diag.), **209**:41
Algae, herbicide biomass effects, **203**:93
Algae, pollution effects, **203**:88
Algal atrazine effects, nutrient absorption, **203**:92
Algal community effects, herbicides, **203**:96
Alkylating agents, cyclophosphamide, **202**:68
Alkylphenol ethoxylate pollution, North America, **207**:37
Alkyphenol ethoxylate residues, Great Lakes (table), **207**:42–43
Alkyphenol ethoxylates sampling locations, Great Lakes (illus.), **207**:39
Allergens, in house dust, **201**:18
Allowable exceedance, conclusion, **209**:75
Alternative methods, for calculating criteria, **209**:37
Ambient air contamination, illicit drugs, **210**:91
Amphibian chronic toxicity, PFOS (perfluorooctanesulfonate), **202**:20
Amphibians, PFOS toxicity, **202**:15
Anaerobic degradation pathway (diag.), **201**:143
Analgesic detections, aquatic environment, **202**:98
Analgesics & anti-inflammatories, description, **202**:63
Analysis of PFOAs, quantification biases, **208**:126
Analysis of PFOSs, quantification biases, **208**:126
Analytical biases, PFOAs (illus.), **208**:127
Analytical methods, perfluoroalkyl isomers, **208**:123

Anemia induction, responsible agents (table), 207:108

Anesthetics, atmospheric fate, 208:13

Anesthetics, chlorine atom and hydroxyl kinetics (table), 208:14

Anesthetics, pulmonary toxins, 201:53

Anesthetics, sources and levels, 208:61

Animal adaptation role, HSP (heat shock protein) production, 206:8

Animal behavior, description, 210:36

Animal behavior, radionuclide effects, 210:35 ff.

Animal exposure to lead, Red Dog Mine, 206:56

Animal residues, PFOA isomers (table), 208:134

Animal residues, PFOS isomers, 208:139

Animal senses, disruption processes, 210:38

Animal waste contamination, chromium, 210:5

Animals, fenamiphos acute toxicity (table), 205:143

Antagonism, toxicity of mixtures, 209:84

Antarctic fish, muted heat-stress response, 206:10

Antibiotics, description, 202:63

Antibiotics, environmental detections, 202:98

Antibiotics, penicillins, 202:64

Antibiotics, quinolones, 202:64

Antibiotics, sulfonamids, 202:64

Antibiotics, tetracyclines, 202:65

Antidepressants, environmental detections, 202:101

Antiepileptics, environmental detections, 202:101

Anti-inflammatory detections, aquatic environment, 202:98

Antimicrobial detections, drinking water, 207:24

Antineoplastic agents, environmental detections, 202:101

Antioxidant enzymes, resist oxidative stress, 206:5

Antioxidant protection, dermal toxicity, 203:119 ff.

Antioxidants, benefit in pulmonary toxicity, 201:59

Antioxidants, oxidative stress mitigation, 206:4

Antioxidants, pollution biomarkers, 206:5

Antioxidants, pulmonary toxicity protection, 201:41 ff.

Antioxidants, toxicity mitigation, 203:131

Anti-pollutant applications, non-thermal plasmas (NTPs) chemistry, 201:120

Apoptosis, HSP effect, 206:14

Applications, PFA isomer profiling, 208:111 ff.

Aquatic criteria development, needed ecotoxicity data, 209:32

Aquatic criteria role, multi-pathway exposure, 209:14

Aquatic criteria, estimation technique, 209:15

Aquatic criteria, exposure-time parameter, 209:14

Aquatic criteria, harmonization with air values, 209:132

Aquatic criteria, harmonization with sediment values, 209:132

Aquatic criteria, species data requirements, 209:8

Aquatic ecotoxicology, *Gammarus* spp., 205:1 ff.

Aquatic environment quality, pesticides, 203:88

Aquatic environment, TFA ubiquity, 208:92

Aquatic field data, evaluation & use, 209:102

Aquatic invertebrates, levels of PFCs (table), 208:189

Aquatic lab data, rating method (tables), 209:27, 209:28

Aquatic life criteria, data requirements (table), 209:7

Aquatic life criteria, ecosystem protection, 209:6

Aquatic life criteria, toxicity data required, 209:7

Aquatic life water quality, pesticides, 209:1 ff.

Aquatic macrophytes, PFOS chronic toxicity, 202:19

Aquatic metabolism, fenoxycarb (table), 202:177

Aquatic organism growth, bioconcentration effects, 204:14

Aquatic organism metabolism, organo-chlorine & -phosphorus pesticides, 204:85

Aquatic organisms, glutathione-S-transferases, 204:75

Aquatic organisms, acyltransferases, 204:79

Aquatic organisms, conjugation reactions (diag.), 204:74

Aquatic organisms, HSPs as biomarkers, 206:17

Aquatic organisms, hydrophobicity effects, 204:8

Aquatic organisms, pesticide bioaccumulation, 204:1 ff.

Aquatic organisms, pesticide bioconcentration, 204:1 ff.

Aquatic organisms, pesticide metabolism, 204:1 ff.

Aquatic organisms, PFFA (perfluorinated fatty acids) ecotoxicity, 202:11

Aquatic outdoor data, evaluation & use, 209:102

Aquatic outdoor field data, quality rating scheme (tables), 209:30–31

Aquatic plants, fluorinated chemical toxicity (diag.), 202:45

Aquatic pollutants, oxidative stress, 206:4

Aquatic species BCF (bioconcentration factor), pesticide clearance times, 204:23

Aquatic species behavioral effects, uranium, 210:46

Aquatic species bioconcentration, humic acid effects, 204:18

Aquatic species bioconcentration, lipid-content effects, 204:9

Aquatic species bioconcentration, organochlorine pesticides (table), 204:24

Aquatic species bioconcentration, organophosphorus pesticides (table), 204:30

Aquatic species bioconcentration, pesticides (table), 204:35

Aquatic species bioconcentration, pyrethroid insecticides (table), 204:34

Aquatic species chronic toxicity, PFOS (table), 202:17

Aquatic species differences, pesticide metabolism, 204:90

Aquatic species effects, coumaphos, 205:152

Aquatic species effects, isofenphos, 205:146

Aquatic species metabolism, glucose & glucuronic acid transferases, 204:72

Aquatic species metabolism, organochlorine pesticides (table), 204:86

Aquatic species metabolism, pesticides (table), 204:67

Aquatic species metabolism, pesticides, 204:89

Aquatic species pesticide metabolism, enzymes, 204:54

Aquatic species pesticide metabolism, esterases, 204:63

Aquatic species toxicity, fenamiphos (table), 205:141

Aquatic species toxicity, fluorotelomer carboxylic acids (FTCA) (diag.), 202:43

Aquatic species, BCF vs. log K_{ow} (diag.), 204:41

Aquatic species, bioconcentration theory, 204:19

Aquatic species, cadmium behavioral effects, 210:48

Aquatic species, chemical metabolism (table), 204:80

Aquatic species, criteria derivation data needs, 209:33

Aquatic species, ecotoxicity testing, 209:9

Aquatic species, exposure data evaluation, 209:102

Aquatic species, lipid content (table), 204:10

Aquatic species, lipid content variability, 204:11

Aquatic species, pesticide bioaccumulation (table), 204:43

Aquatic species, pesticide biota-sediment accumulation factors (table), 204:46

Aquatic species, pesticide metabolic pathways (table), 204:53

Aquatic species, regulatory testing, 204:2

Aquatic system, chemical flow (diag.), 204:3

Aquatic toxicity data, UCDM, 209:98

Aquatic toxicity, PFBS (perfluorobutanesulfonate) (table), 202:23

Aquatic toxicology, perfluorinated chemicals, 202:1 ff.

Aquatic-criteria setting, toxicity data summary (table), 209:21

Arctic avian species, use as bioindicators, 205:105

ARDS (adult respiratory distress syndrome), reactive oxygen species (ROS), 201:43

Aromatic pollutant degradation, ring-hydroxylating oxygenases (RHOs), 206:65 ff.

Asbestos, pulmonary toxin, 201:55

Assessment factor (AF) procedure, criteria derivation, 209:60

Assessment of risk, *Pseudomonas aeruginosa*, 201:72 ff.

Assumptions made, UCDM, 209:93

Assumptions to derived criteria, review, 209:133

Asthma reduction, home visits, 201:25

Asthma, allergens & microbes, 201:18

Asthma, induced by oxidative stress, 201:59

Atmospheric concentrations, FTAcs (table), 208:78

Atmospheric concentrations, FTOHs (table), 208:68

Atmospheric concentrations, FTOs (table), 208:76

Atmospheric concentrations, HFC-134a (table), 208:63

Atmospheric concentrations, NAFSAs (table), 208:83
Atmospheric concentrations, NAFSEs (table), 208:87
Atmospheric contamination, PFAs, 208:1 ff.
Atmospheric degradates, FTAcs, 208:55
Atmospheric degradates, FTALs, 208:46
Atmospheric degradates, FTIs, 208:53
Atmospheric degradates, FTOHs, 208:49
Atmospheric degradates, FTOs, 208:51
Atmospheric degradates, HCFCs, 208:17, 21
Atmospheric degradates, HFOs, 208:30
Atmospheric degradates, NAFSE, 208:60
Atmospheric degradates, oFTOHs, 208:47
Atmospheric degradates, PFALs, 208:36
Atmospheric fate, HCFCs, 208:16
Atmospheric fate, isoflurane, 208:15
Atmospheric fate, perfluoroalkanesulfon-amides, 208:56
Atmospheric formation, PFAs, 208:4
Atmospheric formation, PFCAs, 208:2, 4, 8
Atmospheric levels & sources, saturated HFCs, 208:62
Atmospheric levels, fluorinated anesthetics, 208:61
Atmospheric levels, FTAcs, 208:77
Atmospheric levels, FTOHs & FTOs, 208:67
Atmospheric levels, NAFSA & NAFSE, 208:82
Atmospheric lifetime, FTAc, 208:54
Atmospheric lifetime, FTALs, 208:45
Atmospheric lifetime, FTOHs, 208:48
Atmospheric lifetime, FTOs, 208:51
Atmospheric lifetime, HFC-125, 208:27
Atmospheric lifetime, HFOs, 208:30
Atmospheric lifetime, NAFSE, 208:59
Atmospheric lifetime, oFTOHs, 208:47
Atmospheric lifetime, PFALs, 208:35, 41
Atmospheric lifetime, volatile anesthetics, 208:13
Atmospheric lifetimes, FTIs, 208:53
Atmospheric lifetimes, HCFCs, 208:21
Atmospheric lifetimes, saturated HFCs, 208:22
Atmospheric oxidation, perfluorinated radical mechanism, 208:5
Atmospheric oxidation, perfluoroacyl halides (diag.), 208:8
Atmospheric perfluorinated acids chemistry, 208:1 ff.
Atmospheric sources & levels, HCFCs, 208:61
Atmospheric sources, fluorinated anesthetics, 208:61

Atmospheric sources, fluorotelomer compounds, 208:66
Atmospheric sources, perfluorosulfonamides, 208:77
Atrazine effects, algal nutrient absorption, 203:93
Atrazine effects, diatom species composition, 203:94
Atrazine toxicity data, comparative distribution fit (illus.), 209:49
Atrazine, data-set distribution test (diag.), 209:42
Auditory loss, human aging, 207:113
Averaging periods, criteria derivation, 209:69
Avian exposure parameters, surrogate species (table), 202:37
Avian predators, TRV (toxicity reference values) for PFBS (table), 202:37
Avian species effects, coumaphos, 205:152
Avian species effects, isofenphos, 205:146
Avian species residues, PBDE, 207:63
Avian species, bioindicators, 205:105
Avian water quality criteria, PFOS, 202:33
Azinphos-methyl frequently detected, Great Lakes waters, 207:19

B
B_2-Sympathomimetics, environmental detections, 202:102
Babies and ADHD, pollutants, 201:2
Babies, house dust exposure, 201:1
Bacteria, free-living pseudomonads, 201:72
Bacteria, in house dust, 201:18
Bacterial bioremediation, chromium, 210:15
Bacterial cell signaling, chemical exposure, 204:140
Bacterial growth rate, formula, 203:9
Bacterial infections, nosocomial pneumonia (table), 201:75
BAF (Bioaccumulation factor), formula, 203:14
BAF (bioaccumulation), controlling factors, 204:42
Baltic Sea, illegal oil discharges (illus.), 206:109
Baltic Sea, oil spills, 206:107
Baltic Sea, shipping accident types (table), 206:108
Baltic Sea, shipping accidents & oil spills (table), 206:107
Baltic Sea, shipping accidents, 206:106
Batie classes, RHOs, 206:67
Batie classification scheme, RHOs (table), 206:69

Batie classification, nature & deficiencies, 206:68

BCF (bioconcentration factor), defined, 204:4

BCF (bioconcentration factors), PFOS (table), 202:9

BCF effect, pesticide class, 204:42

BCF effects, pesticide clearance times, 204:23

BCF for fish, natural vs. lab, 202:10

BCF vs. log K_{ow}, pesticides in aquatic species (diag.), 204:41

BCF vs. log K_{ow}, pesticides in fish (diag.), 204:7

BCF, PFOS (table), 202:9

BCF, physico-chemical effects (table), 204:5

Behavior in animals, radionuclide effects, 210:35 ff.

Behavior testing, drift and foraging activity, 205:29

Behavioral assays, gammarids, 205:22

Behavioral control processes, ecological effects (diag.), 210:38

Behavioral effects, cobalt irradiation & exposure, 210:48

Behavioral effects, formaldehyde, 203:111

Behavioral link, physiological mechanisms, 210:37

Behavioral responses, metals, 210:42

Behavioral studies, fish & wildlife, 210:45

Behavioral test methods, gammarid toxicity (table), 205:23

Benzene, pulmonary toxin, 201:49

Beta (β) Subunit structure, RHOs, 206:73

Beta-blockers, description, 202:65

Beta-blockers, environmental detections, 202:99

Bioaccumulation effects, dissolved organic carbon, 204:47

Bioaccumulation factor (BAF), controlling factors, 204:42

Bioaccumulation factor (BAF), formula, 203:14

Bioaccumulation factors, PCB effects (table), 201:51

Bioaccumulation in gulls, feeding affects, 205:87

Bioaccumulation of pesticides, aquatic species (table), 204:43

Bioaccumulation of pesticides, theory, 204:49

Bioaccumulation, description, 204:2

Bioaccumulation, dietary route, 204:47

Bioaccumulation, ecological effects, 204:45

Bioaccumulation, flow in aquatic system (diag.), 204:3

Bioaccumulation, influential factors, 205:86

Bioaccumulation, metabolic effects, 204:48

Bioaccumulation, PCBs, 201:147–148

Bioaccumulation, pesticides, 204:1 ff.

Bioaccumulation, PFA isomers, 208:149

Bioaccumulation, PFOS, 202:8

Bioaccumulation, role in criteria setting, 209:127

Bioaccumulation, sediment effects, 204:44

Bioaccumulation, UCDM & food residues, 209:89

Bioaccumulation, vs. bioavailability & biodegradation, 203:3

Bio-active agents, electron transfer functionalities, 201:42

Bioactive α-dicarbonyls, examples, 204:143

Bioassay verification, bioavailability (table), 203:43

Bioassays for bioavailability, plants (table), 203:36

Bioassays in soil fauna, bioavailability estimation (table), 203:38

Bioassays, measuring bioavailability, 203:29

Bioassays, microbial bioavailability (table), 203:31

Bioaugmentation, chromium bioremediation, 210:14

Bioavailability bioassays, soil fauna (table), 203:38

Bioavailability connection, Monod equation, 203:47

Bioavailability effect, criteria compliance, 209:118

Bioavailability effect, mycorrhizal fungi, 203:12

Bioavailability estimates, octanol-water partitioning, 203:45

Bioavailability estimates, simulating modeling, 203:46

Bioavailability estimation method, chemicals (table), 203:63

Bioavailability estimation, chemical extraction (table), 203:43

Bioavailability estimation, higher plants (table), 203:36

Bioavailability estimation, models, 203:47, 51

Bioavailability evaluation, methods (table), 203:61

Bioavailability from soil, OP pesticides, 205:129

Bioavailability in plants, uptake models, 203:51

Bioavailability measurement, chemical
   extraction, **203**:41
Bioavailability models, soil fauna, **203**:57
Bioavailability of chemicals, influencing
   factors (table), **203**:6
Bioavailability of chemicals, microbial
   bioassays (table), **203**:31
Bioavailability of hydrocarbons, marine
   species, **206**:106
Bioavailability of OP pesticides, earthworms,
   205:129
Bioavailability of OP pesticides, microbes,
   205:129
Bioavailability testing, mining wastes, **206**:34
Bioavailability verification, bioassays (table),
   203:43
Bioavailability, bioassay measurement, **203**:29
Bioavailability, bound-residue mobilization,
   203:26
Bioavailability, criteria-setting implications,
   209:76
Bioavailability, definition, **203**:2
Bioavailability, dynamic processes, **203**:4
Bioavailability, environmental effects, **203**:27
Bioavailability, equation symbols (table),
   203:65
Bioavailability, genotoxicity bioassay, **203**:33
Bioavailability, metals, **203**:4
Bioavailability, microbial models, **203**:47
Bioavailability, organic chemicals, **203**:4
Bioavailability, surfactant effects (table),
   203:22
Bioavailability, temperature & moisture
   effects, **203**:27
Bioavailability, vs. bioaccumulation
   & biodegradation, **203**:3
Bioavailability, vs. toxicokinetics, **203**:4
Bioavailability, xenobiotics in soil, **203**:1 ff.
Bioconcentrated pesticides, organ distribution,
   204:15
Bioconcentration effects, biological factors,
   204:14
Bioconcentration effects, humic acids, **204**:18
Bioconcentration factor (BCF), defined, **204**:4
Bioconcentration in aquatic species,
   organochlorine pesticides (table),
   204:24
Bioconcentration in aquatic species,
   organophosphorus pesticides (table),
   204:30
Bioconcentration in aquatic species, pesticides
   (table), **204**:35

Bioconcentration in aquatic species, pyrethroid
   insecticides (table), **204**:34
Bioconcentration of PFOS, fish, **202**:10
Bioconcentration potential, test regulations,
   204:2
Bioconcentration theory, absorption
   & elimination, **204**:19
Bioconcentration, controlling factors, **204**:3
Bioconcentration, description, **204**:2
Bioconcentration, dissolved organic carbon
   effects, **204**:18
Bioconcentration, environmental effects,
   204:16
Bioconcentration, flow in aquatic system
   (diag.), **204**:3
Bioconcentration, illumination effects, **204**:16
Bioconcentration, isomerism effects, **204**:9
Bioconcentration, lipid-content effects, **204**:9
Bioconcentration, metabolism effects, **204**:14
Bioconcentration, PCBs, **201**:147–148
Bioconcentration, pesticides, **204**:1 ff.
Bioconcentration, PFOS, **202**:8
Bioconcentration, pH-dependency, **204**:17
Bioconcentration, theory, **204**:19
Biodegradability, low-fluorine organics,
   208:171
Biodegradability, pharmaceuticals (table),
   202:120–123
Biodegradation of oil spills, marine
   environment, **206**:103
Biodegradation pathway in soil, 8:2-FTOH
   (diag.), **208**:167
Biodegradation pathway, N-EtFOSE (diag.),
   208:169
Biodegradation pathway, TFMP-NS (diag.),
   208:173
Biodegradation pathways, TFM-DS (diag.),
   208:172
Biodegradation, fluorinated alkyl substances,
   208:161 ff.
Biodegradation, fluorinated polymers, **208**:170
Biodegradation, fluorotelomer-bases
   chemicals, **208**:165
Biodegradation, FTOHs, **208**:165
Biodegradation, heroin, **210**:80
Biodegradation, PCBs, **201**:142
Biodegradation, PFOS derivatives, **208**:168
Biodegradation, PFOS, **202**:6
Biodegradation, vs. bioavailability
   & bioaccumulation, **203**:3–4
Bioenergetic responses, gammarids, **205**:32
Biofilms, *Pseudomonas aeruginosa*, **201**:92

Bioindicator species in the European Arctic, Svalbard glaucous gull, **205**:77 ff.

Bioindicator species, Arctic birds, **205**:105

Biological aspects, San Francisco Bay water quality, **206**:130

Biological degradation, OP pesticides, **205**:127

Biological factors, bioconcentration effects, **204**:14

Bioluminescence assay, naphthalene, **203**:33

Biomagnification factor (BMF), default values (table), **209**:91

Biomagnification, description, **204**:2

Biomagnification, PCBs, **201**:147, 149

Biomagnification, PFOS, **208**:194

Biomarker for detoxification, glutathione-S-transferase, **205**:52

Biomarker tests, gammarids, **205**:32

Biomarker, acetylcholinesterase, **205**:50

Biomarker, HSPs in fish, **206**:17

Biomarkers in gammarids, vitellogenin-like proteins, **205**:45

Biomarkers of pollution, antioxidants, **206**:5

Biomarkers of stress, aquatic contaminants, **206**:4

Biomarkers, in *Gammarus* spp. (table), **205**:38

Biomarkers, metallothioneins and lipid peroxidation, **205**:49

Biomarkers, Svalbard glaucous galls, **205**:89

Biomass effects in algae, herbicides, **203**:93

Biomonitoring system, multispecies freshwater biomonitor (MFB), **205**:27

Biomonitors, fish, **206**:2

Bioremediation technology, chromium, **210**:13

Bioremediation, chromium toxicity attenuation, **210**:1 ff.

Bioremediation, pollution removal, **210**:2

Biostimulation, chromium bioremediation, **210**:14

Biosurfactant exudate, microbes, **203**:25

Biota & sediment residue, PBDE (poly-brominated diphenyl ethers; table), **207**:67

Biota residues, alkyphenol ethoxylates (table), **207**:43

Biota residues, chlorinated paraffins (table), **207**:82

Biota residues, flame retardants (table), **207**:76

Biota residues, HBCD (hexabromocyclode-cane), **207**:72

Biota residues, perfluorinated compounds, **207**:54

Biota residues, synthetic musks (table), **207**:50

Biota sampling locations, perfluorinated surfactants (illus.), **207**:56

Biota-sediment accumulation factors (BSAF) for pesticides, aquatic species (table), **204**:46

Biota-soil accumulation factor (BSAF), formula, **203**:13

Bio-uptake by aquatic species, metabolic effects, **204**:51

Bird behavioral effects, caesium, **210**:46

Bird contamination, San Francisco Bay, **206**:123

Birds & water quality, San Francisco Bay, **206**:137

Birds, effects of coumaphos, **205**:152

Birds, effects of isofenphos, **205**:146

Bisphenol A (BPA) residues, Great Lakes, **207**:33–34

Bleomycin, pulmonary toxin, **201**:53

Blood-brain barrier effects, aging, **207**:111

Blood-brain barrier function, interfering chemicals (table), **207**:112

Bluegill cumulative mortality, PFOS (table), **202**:29

B-lymphocyte function, aging effects, **207**:118

BMF (biomagnification factor), default values (table), **209**:91

Bone density decrease, aging effects, **207**:108

Bound residue, definition, **203**:26

Bound residues, desorption from soil, **203**:20

Bound residues, soil aging effects, **203**:25

Bound residues, xenobiotic soil occlusion, **203**:26

Bound-residue mobilization, bioavailability, **203**:26

Brain effects, uranium, **210**:39

Brain volume, aging effect, **207**:110

Breakdown in water, coumaphos, **205**:151

Brominated diphenyl ethers (BDE), San Francisco Estuary (illus.), **206**:127

Brominated flame retardants, Svalbard glaucous galls, **205**:81

Bronchiolitis oblitherans, lung disease, **204**:133

BSAF (Biota-soil accumulation factor), formula, **203**:13

Burkholderia, human pathogenicity, **201**:74

Burn-wound infections, *Pseudomonas aeruginosa*, **201**:80

Burr III distribution fit, pesticide toxicity data (table), **209**:50

Burr Type III distribution fit, pesticide data sets (table), **209**:51

C

Cadmium behavioral effects, aquatic species, **210**:48

Cadmium behavioral effects, invertebrates, **210**:47

Cadmium exposure, humans, **206**:51

Cadmium, human behavior response, **210**:43

Cadmium, rodent behavioral studies, **210**:45

Caesium behavioral effects, Chernobyl wildlife, **210**:47

Caesium, bird behavioral effects, **210**:46

Caesium, human behavioral response, **210**:43

Caesium, rodent behavioral studies, **210**:45

Canadian quality guidelines, sediment, **207**:46

Cancer, PAHs (polyaromatic hydrocarbons) and house dust, **201**:11

Carbamate insecticides, neurotoxic biomarker, **205**:50

Carbamate pesticides, chemical structures (illus.), **204**:102

Carbon disulfide, pulmonary toxin, **201**:51

Carbon monoxide, pulmonary toxin, **201**:47

Carbon tetrachloride, pulmonary toxin, **201**:50

Carcinogen in humans, chromium, **210**:2

Carcinogenicity, formaldehyde, **203**:107

Carcinogenicity, isofenphos, **205**:148

Cardiovascular system, aging effects, **207**:102

Carpet contamination, lead, **201**:4

Carpet contamination, PAHs & PCBs, **201**:8

Carpet contamination, pesticides, **201**:6

Carpet contamination, vacuum cleaning, **201**:21

Carpet vs. uncovered floors, contamination and cleaning, **201**:20

Carvoxime, toxicity, **203**:127

Catalytic pockets, RHO active sites (illus.), **206**:83–84

Cell sensitivity, toxicity in diatoms, **203**:89

Cell signaling, chemical exposure, **204**:140

Cell signaling, dermal toxicity, **203**:119 ff., 131.

Cell signaling, electron transfer & reactive oxygen species, **204**:140

Cellular injury, ROS, **201**:43

Cephalosporins, antibiotic, **202**:65

Characteristics, fluorinated alkyl substances (table), **208**:163

Characteristics, fluorinated polymers, **208**:170

Chelate-assisted uptake, phytoextraction, **210**:20

Chemical accumulation, earthworms, **203**:35

Chemical adsorption effects, dissolved organic matter, **203**:21

Chemical adsorption effects, surfactants, **203**:21

Chemical and biological warfare agents (CBW), disposal by NTPs chemistry, **201**:128

Chemical aspects, San Francisco Bay water quality, **206**:130

Chemical bioavailability, higher plants (table), **203**:36

Chemical bioavailability, influencing factors (table), **203**:6

Chemical bioavailability, microbial bioassays (table), **203**:31

Chemical bioavailability, soil fauna (table), **203**:38

Chemical bioavailability, surfactant effects (table), **203**:22

Chemical classes, OP pesticides, **205**:119

Chemical contaminants, house dust (table), **201**:9

Chemical contaminants, infant exposure, **201**:2

Chemical contaminants, south San Francisco Bay, **206**:120

Chemical contamination, European Arctic, **205**:78

Chemical degradation, microorganisms, **203**:29

Chemical diffusion into roots, formula, **203**:10

Chemical effects, blood-brain barrier function (table), **207**:112

Chemical effects, diatom cell wall, **203**:90

Chemical effects, diatom cytoskeleton, **203**:89

Chemical effects, iodide uptake (table), **207**:123

Chemical exposure, cell signaling effects, **204**:140

Chemical exposure, soil ingestion, **203**:45

Chemical extraction, bioavailability measurement, **203**:41

Chemical extraction, estimating bioavailability (table), **203**:43

Chemical flow, aquatic system (diag.), **204**:3

Chemical hydrophobicity, BCF effects, **204**:8

Chemical mechanisms, non-thermal plasmas chemistry, **201**:119

Chemical metabolism, aquatic species (table), **204**:80

Chemical mixture effects, criteria compliance, **209**:119

Chemical mixtures, criteria-setting implications, **209**:80

Chemical nature, PFOS, **202**:5

Chemical penetration of skin, aging effect,
    207:129
Chemical pollutants, HSP production, 206:7
Chemical pollutants, organic wastewater,
    207:30
Chemical properties, chromium (table), 210:3
Chemical properties, fenamiphos (table),
    205:134
Chemical properties, PFOS (table), 202:6
Chemical stressors, in fish, 206:3
Chemical structures, electron transfer agents
    (illus.), 203:121
Chemical structures, JHAs (juvenile hormone
    agonists) (diag.), 202:158
Chemical structures, juvenile hormones (diag.),
    202:157
Chemical toxicity, freshwater invertebrates and
    fish (table), 205:9
Chemical toxicity, gammarids, 205:7
Chemical transport, phagocytosis, 203:7
Chemical transport, through membranes, 203:5
Chemical uptake by roots, influencing factors,
    203:11
Chemical uptake in plants, PLANTX model,
    203:52
Chemical uptake models, plants, 203:51
Chemical uptake, aquatic organism growth
    effects, 204:14
Chemical uptake, higher plants, 203:34
Chemical uptake, hydrophobic chemicals,
    203:7
Chemical uptake, into plant roots, 203:9
Chemical uptake, mechanisms, 203:7
Chemical uptake, microbes, 203:9
Chemical uptake, soil fauna, 203:12
Chemical-drug interactions, pharmacology,
    207:131
Chemicals (various), structures (illus.), 204:99
Chemicals and soil, interactions, 203:15
Chemicals in soil, determining bioavailability
    (table), 203:63
Chemicals listed in this volume, chemical
    names (table), 209:128
Chemicals monitored in Great Lakes,
    categories (table), 207:5
Chemicals of emerging concern, available
    information (table), 207:5
Chemicals of emerging concern, Great Lakes,
    207:1 ff.
Chemicals of emerging concern, International
    Joint Commission, 207:2
Chemicals, active transport, 203:7

Chemicals, bioavailability evaluation methods
    (table), 203:61
Chemicals, environmental flow (diag.), 203:3
Chemicals, facilitated transport, 203:6
Chemicals, Great Lakes contamination,
    207:1 ff.
Chemicals, physico-chemical effects on BCF
    (table), 204:5
Chemicals, transport mechanisms, 203:5
Chemistry, fenoxycarb, 202:155 ff.
Chemistry, fenoxycarb, 202:159
Chemistry, OP pesticides, 205:118
Chemistry, PFA precursors, 208:12
Chernobyl wildlife, caesium behavioral effects,
    210:47
Children home exposure, pesticide volatility,
    201:7
Children, blood lead levels, 201:6
Children, lead exposure, 206:56
Children, Pseudomonas aeruginosa infection,
    201:83
Children's environmental exposure, hygiene
    hypothesis, 201:19
Chiral legacy organochlorines, glaucous gulls,
    205:81
Chitin and molting, gammarid biomarker,
    205:46
Chlordane toxicity data, comparative
    distribution fit (illus.), 209:47
Chlordane, data-set distribution test (diag.),
    209:42
Chlordane, non-Hodgkins lymphoma, 201:8
Chlorinated paraffin residues, water
    & sediment (table), 207:81
Chlorinated paraffins, environmental
    contamination, 207:78
Chlorine, pulmonary toxin, 201:48
Chlorine-atom kinetics, volatile anesthetics
    (table), 208:14
Chlorofluorocarbon replacements, PFAs, 208:2
Chloroform, pulmonary toxin, 201:50
Chlorpyrifos aquatic criteria, calculated with
    UCDM (table), 209:137
Chlorpyrifos toxicity data, comparative
    distribution fit (illus.), 209:43
Chlorpyrifos, AChE inhibition, 209:11
Chlorpyrifos, data-set distribution test (diag.),
    209:41
Chlorpyrifos, water quality parameters, 209:3
Chromatograms, perfluoroalkyl sulfonates
    (illus.), 208:117
Chromium accumulation, plants, 210:5

Chromium bioremediation, bioaugmentation, 210:14

Chromium bioremediation, biostimulation, 210:14

Chromium bioremediation, yeast & filamentous fungi, 210:15

Chromium complexes, role in transport, 210:6

Chromium contamination, fertilizer, animal waste & sewage sludge, 210:5

Chromium effects, human health, 210:8

Chromium effects, human inhalation, 210:8

Chromium effects, human skin, 210:8

Chromium effects, microorganisms, 210:7

Chromium inhibition, seedling germination & growth, 210:9

Chromium levels, environmental media (table), 210:4

Chromium release, from coal, 210:5

Chromium species, oxidation states (table), 210:3

Chromium toxicity attenuation, by bioremediation, 210:1 ff.

Chromium toxicity, reproduction, 210:8

Chromium transport, plants, 210:5

Chromium, bacterial bioremediation, 210:15

Chromium, bioremediation technology, 210:13

Chromium, chemistry, 210:2

Chromium, electron transport effects (diag.), 210:12

Chromium, environmental levels (table), 210:4

Chromium, environmental sources, 210:4

Chromium, ex & in situ site remediation, 210:17

Chromium, human carcinogen, 210:2

Chromium, macromolecular damage, 210:12

Chromium, microbial remediation, 210:14

Chromium, misc. phytotoxic effects, 210:12

Chromium, oxidation states, 210:2

Chromium, photosynthesis inhibition, 210:10

Chromium, physico-chemical properties (table), 210:3

Chromium, phyto- (green)-remediation, 210:16

Chromium, phytotoxicity, 210:9

Chromium, plant growth retardation, 210:10

Chromium, production & uses, 210:4

Chromium, toxic environmental pollutant, 210:1

Chromium, toxicity, 210:7

Chromium, uses and production, 210:5

Chromium-induced oxidative stress, plants, 210:11

Chronic criteria derivation, herbicides, 209:117

Chronic criterion derivation, SSD procedure, 209:114

Chronic criterion derivation, using ACRs, 209:114

Chronic effects, formaldehyde, 203:108

Chronic infant health, pollutants, 201:2

Chronic medical conditions, the elderly, 207:96

Chronic toxicity estimates, from acute data, 209:16

Chronic toxicity in amphibians, PFOS, 202:20

Chronic toxicity in fish, PFOS, 202:21

Chronic toxicity, fenamiphos, 205:143

Chronic toxicity, isofenphos, 205:147

Chronic toxicity, PFOS in invertebrates, 202:20

Chronic-data gaps, estimation technique, 209:103

Chronology (literature), illicit drugs in the environment (table), 210:63–66

Clandestine drug labs (clan labs), environmental contamination, 210:91

Classification of pseudomonads, genera (table), 201:73

Classification systems, RHOs, 206:66

Classroom pollutants, dust, 201:5

Cleaning carpets vs. floors, contamination, 201:20

Cleaning products, safety, 201:23

Coal burning, PAHs in house dust, 201:13

Coal contamination, chromium, 210:5

Coastal sediment toxicity, G. locusta, 205:8

Cobalt irradiation & exposure, behavioral effects, 210:48

Cobalt, human behavioral response, 210:43

Cocaine, pulmonary toxin, 201:56

Colchicine, diatom toxicity, 203:89

Concentration addition model, toxicity of mixtures, 209:81

Conjugation reactions, aquatic organisms (diag.), 204:74

Consumption, petroleum products (table), 206:96

Contaminant effects, Svalbard glaucous galls (table), 205:89

Contaminant erosion, buried sediments, 206:131

Contaminant exposure, thermoregulatory effects, 205:97

Contaminant genotoxicity, Svalbard glaucous galls, 205:100

Contaminant levels and patterns, Svalbard glaucous galls, 205:79

Contaminant reduction, home visits, 201:25

Contaminant regulation, among countries
    (table), **207**:7–11
Contaminant removal, phytoremediation,
    206:87
Contaminant sequestration in marshes, San
    Francisco Bay, **206**:137
Contaminants research, Svalbard glaucous
    galls, **205**:77 ff.
Contaminants vs. nest temperature, Svalbard
    glaucous galls (illus.), **205**:98
Contaminants, illicit drugs in the environment,
    210:59 ff.
Contaminants, in house dust, **201**:3
Contaminants, iron-ore tailings, **206**:32
Contaminants, Svalbard glaucous gall residues,
    205:84
Contaminate effects, on avian species, **205**:104
Contaminate list, definitions (table), **205**:109
Contaminated money supply, illicit drugs,
    210:90
Contaminated sediment toxicity, *G. pulex*,
    205:8
Contaminated soil contact, Red Dog Mine,
    206:54
Contamination & stress, biomarkers, **206**:4
Contamination by metals, classroom dust
    (table), **201**:5
Contamination of air, illicit drugs, **210**:91
Contamination prevention, hand washing,
    201:23
Contamination, San Francisco Bay, **206**:124
Contamination, south San Francisco Bay,
    206:120
Contamination, vacuum cleaning, **201**:20
Contrast media, environmental detections,
    202:102
Contrasts as environmental contaminants, licit
    vs. illicit drugs, **210**:73
COPD (chronic obstructive pulmonary
    disease), relative to oxidative stress,
    201:59
Coumaphos, aquatic species effects, **205**:152
Coumaphos, avian species effects, **205**:152
Coumaphos, description, **205**:150
Coumaphos, ecotoxicology, **205**:152
Coumaphos, environmental fate, **205**:150
Coumaphos, fate in water, **205**:151
Criteria calculation, approaches, **209**:37
Criteria calculation, SSD method, **209**:38
Criteria compliance, bioavailability effect,
    209:118
Criteria compliance, incorporating water
    quality effects, **209**:117

Criteria compliance, mixture effects, **209**:119
Criteria compliance, temperature & pH effects,
    209:121
Criteria derivation data needs, aquatic species,
    209:33
Criteria derivation flow chart, UCDM (diag.),
    209:97
Criteria derivation improvement, data
    generation, **209**:94
Criteria derivation methods, evaluating
    ecotoxicity data, **209**:25
Criteria derivation, AF procedure, **209**:60
Criteria derivation, averaging periods, **209**:69
Criteria derivation, data outliers, **209**:54
Criteria derivation, data reduction methods,
    209:35
Criteria derivation, data requirements (table),
    209:7
Criteria derivation, field data role, **209**:13
Criteria derivation, national water quality,
    209:2
Criteria derivation, rating single-species data
    quality (table), **209**:26
Criteria derivation, role of ACRs, **209**:66
Criteria derivation, SSD percentile cutoff point,
    209:52
Criteria derivation, taxa aggregation, **209**:54
Criteria derivation, using default ACRs,
    209:116
Criteria derivation, using multispecies, **209**:13
Criteria derivation, water quality effects,
    209:76
Criteria setting role, bioaccumulation, **209**:127
Criteria setting role, secondary toxicity,
    209:127
Criteria setting, 5th percentile calculation,
    209:112
Criteria setting, role of ecosystem data,
    209:125
Criteria setting, role of TES, **209**:125
Criteria setting, sediment harmonization,
    209:92
Criteria setting, sensitive species protection,
    209:124
Criteria setting, SSD flow chart (diag.),
    209:111
Criteria validation, against ecotoxicity data,
    209:124
Criteria-setting implications, bioavailability,
    209:76
Criteria-setting, chemical mixtures, **209**:80
Criteria-setting, ecotoxicity data checking,
    209:88

Cross (or protection) tolerance, HSP, **206**:11

Crude oil, constituents, **206**:98

Culturing gammarids, methods, **205**:6

Current production, PFOS, **208**:121

Current-use pesticide detections, sediments (table), **207**:22

Current-use pesticide detections, US streams (table), **207**:20–21

Current-use pesticide pollution, Great Lakes (table), **207**:16

Cyclophosphamide, alkylating agent, **202**:68

Cystic fibrosis, *Pseudomonas aeruginosa*, **201**:84

Cytochrome P450, reaction scheme (diag.), **204**:54

Cytokine function, from aging, **207**:118

D

Daphnia toxicity, PFOS, **202**:14

Daphnia, post-exposure feeding depression, **205**:20

Data collection details, UCDM, **209**:98

Data flow chart, UCDM (diag.), **209**:97

Data generation, criteria derivation improvement, **209**:94

Data outliers, in criteria derivation, **209**:54

Data quality, ecotoxicity data summaries, **209**:20

Data quantity required, ecotoxicity, **209**:32

Data reduction methods, criteria derivation, **209**:35

Data relevance & reliability scores, data categories (table), **209**:31

Data requirements, criteria derivation (table), **209**:7

Data requirements, UCDM, **209**:6

Data sources, for UCDM development (table), **209**:17

Data-gap filling, estimation techniques, **209**:15

DDT contamination, European Arctic, **205**:80

DDT toxicity data, comparative distribution fit (illus.), **209**:43

DDT, data-set distribution test (diag.), **209**:41

DEET detections, Great Lakes waters, **207**:36

Default ACR calculation, pesticides (table), **209**:69

Definitions, UCDM, **209**:95

Degradability, PFOS, **208**:165

Degradation behavior, fluorotelomer ethoxylates, **208**:168

Degradation improvements, RHOs, **206**:85

Degradation in soil and water, fenamiphos, **205**:134

Degradation models, microbial bioavailability, **203**:47

Degradation of isofenphos, vegetation, **205**:146

Degradation of pollutants, oxygenases, **206**:65 ff.

Degradation product structures, fenoxycarb (diag.), **202**:161

Depuration, gammarid toxicity, **205**:19

Dermal & general toxicity, TCE (trichloroethylene), methyl bromide & geraniol, **203**:128

Dermal contact & transfer, illicit drugs, **210**:93

Dermal effects, UV radiation & metals, **203**:123

Dermal toxicity, antioxidant protection, **203**:119 ff.

Dermal toxicity, cell signaling, **203**:131

Dermal toxicity, electron transfer & reactive oxygen species, **203**:119 ff.

Dermal toxicity, linalool, sunscreens & fullerenes, **203**:129

Dermal toxicity, methyl salicylate & methotrexate, **203**:130

Dermal toxicity, oxidative stress & cell signaling, **203**:119 ff.

Dermal toxicity, oxygen-mediated effects, **203**:120

Dermatotoxic mechanisms, electron transfer and reactive oxygen species, **203**:121

Dermatotoxicity schematic, reactive oxygen species (diag.), **203**:122

Dermatotoxicity, mustard gas, acrylamide & dioxins, **203**:126

Designer drugs, general public exposure, **210**:82

Desorption of PFOS, soil (table), **202**:7

Detection frequency, pharmaceuticals in water, **207**:27

Developmental role in fish, HSPs, **206**:10

Diacetyl & other α-dicarbonyls, protein glycation, **204**:142

Diacetyl connection, ethanol metabolites, **204**:140

Diacetyl mechanism of action, electron transfer, **204**:139

Diacetyl toxic effects, metabolite vs. parent, **204**:136

Diacetyl toxicity, caused by electron transfer, **204**:133 ff.

Diacetyl toxicity, lung effects, **204**:134

Diacetyl toxicity, oxidative effects, **204**:134

Diacetyl, electron affinity, **204**:138

Diacetyl, enzyme activator, **204**:145

Diacetyl, genotoxicity, **204**:144

Diacetyl, mechanism of action, **204**:136, **204**:139

Diacetyl, molecular electrostatic potential, **204**:139

Diacetyl, pulmonary toxin, **201**:52

Diacetyl, ROS (reactive oxygen species) production, **204**:135

Diacetyl, structure (illus.), **204**:137

Diatom cell-density effects, herbicides, **203**:94

Diatom community effects, herbicides, **203**:96

Diatom cytoskeleton, herbicide effects, **203**:89

Diatom growth effects, herbicides, **203**:93

Diatom nucleus, toxic effects, **203**:89

Diatom pesticide sensitivity, interfering factors, **203**:96

Diatom sensitivity, cytology & ultrastructure, **203**:89

Diatom species composition, atrazine effects, **203**:94

Diatom species effects, herbicides, **203**:94

Diatoms, pesticide effects, **203**:87 ff.

Diatoms, siliceous cell wall effects, **203**:90

Diazinon frequently detected, Great Lakes waters, **207**:19

Diazinon toxicity data, comparative distribution fit (illus.), **209**:48

Diazinon, data-set distribution test (diag.), **209**:42

Dieldrin toxicity data, comparative distribution fit (illus.), **209**:46

Dieldrin, data-set distribution test (diag.), **209**:42

Dietary route, bioaccumulation, **204**:47

Dioxin contamination, San Francisco Bay, **206**:127

Dioxins, toxicity, **203**:126

Direct food contamination, PFCs, **208**:196

Disease in infants, pollutants, **201**:2

Disease transmission, *Pseudomonas aeruginosa*, **201**:97

Disease-free water, *Pseudomonas aeruginosa* (table), **201**:91

Disinfectant residues, surface waters (table), **207**:31

Disinfection, *Pseudomonas aeruginosa*, **201**:96

Dispersion of oil spills, marine environment, **206**:102

Disruption processes, animal senses, **210**:38

Dissolved organic carbon, bioaccumulation effects, **204**:47

Dissolved organic carbon, bioconcentration effects, **204**:18

Distribution, PCBs, **201**:137 ff.

Diversion, illicit drugs, **210**:93

DNA effects in diatoms, chemicals, **203**:89

Dose reconstruction, illicit drug source assessment, **210**:82

Doxorubicin, toxicity, **203**:129

Drift behavior and pollution, *Gammarus*, **205**:22

Drift behavior and predators, *Gammarus*, **205**:22

Drinking water cleaning, pharmaceuticals (table), **202**:112–117

Drinking water contamination, *Pseudomonas aeruginosa*, **201**:87

Drinking water detections, pharmaceuticals (table), **202**:90–95

Drinking water residues, antimicrobials, **207**:24

Drinking water residues, PFCs (table), **208**:200

Drinking water risks, *Pseudomonas aeruginosa* (table), **201**:104

Drinking water, PFC contamination, **208**:199

Drug adulterants, potential environmental contaminants, **210**:81

Drug consumption level, the US elderly, **207**:132

Drug contaminants, environment, **210**:59 ff.

Drug disposal, pharmaceutical release, **202**:73

Drug impurities, potential environmental contaminants, **210**:81

Drug interactions, hospital admissions, **207**:133

Drugs of abuse, detected in environmental media (table), **210**:77–79

Drugs of abuse, frequent detects (table), **210**:75

Dust and metal regulation, Red Dog Mine, **206**:60

Dust control, Red Dog Mine, **206**:60

Dust exposure, Red Dog Mine, **206**:50

Dust mites, in house dust, **201**:18

Dust-bearing soils, lead exposure (illus.), **206**:55

Dust-borne metal exposure, Red Dog Mine, **206**:53

E

Ear infections, *Pseudomonas aeruginosa*, **201**:81

Earthworm, chemical uptake, **203**:35

Earthworms, bioavailability of OP pesticides, **205**:129

Earthworms, chemical bioavailability (table), **203**:38

Earthworms, chemical uptake, **203**:12

ECF (electrochemical fluorination) PFOA, isomer profile (table), **208**:132

Ecological biomarkers, Svalbard glaucous galls, **205**:89

Ecological consequences, radionuclide effects, **210**:49

Ecological effects, behavioral control processes (diag.), **210**:38

Ecological factors, bioaccumulation effects, **204**:45

Ecosystem data, role in criteria setting, **209**:125

Ecosystem protection, aquatic life criteria, **209**:6

Ecotoxicity data checking, criteria setting, **209**:88

Ecotoxicity data evaluation & use, single- & multi-species, **209**:101

Ecotoxicity data evaluation, criteria derivation methods, **209**:25

Ecotoxicity data evaluation, UCDM, **209**:101

Ecotoxicity data required, UCDM, **209**:99

Ecotoxicity data summaries, quality ratings, **209**:20

Ecotoxicity data, criteria validation, **209**:124

Ecotoxicity data, quantity required, **209**:32

Ecotoxicity evaluation, hypothesis tests vs. regression analysis, **209**:9

Ecotoxicity, hypothesis testing issues, **209**:9

Ecotoxicity, illicit drugs, **210**:94

Ecotoxicity, PFFAs, **202**:11

Ecotoxicity, regression analysis, **209**:10

Ecotoxicity, vs. bioavailability, **203**:4

Ecotoxicology studies, pharmaceuticals, **202**:54

Ecotoxicology, coumaphos, **205**:152

Ecotoxicology, fenamiphos, **205**:139

Ecotoxicology, *Gammarus* spp., **205**:1 ff.

Ecotoxicology, isofenphos, **205**:146

EDCs (endocrine disrupting chemicals), house dust, **201**:14, 17

Edible fish residues, long-chain perfluorinated substances (table), **208**:187

Edible fish residues, PFCs (table), **208**:181

Edible invertebrates, PFC levels, **208**:193

Effluents from mining, characteristics, **206**:30

Elderly people, chronic medical conditions, **207**:96

Elderly people, sensitive populations, **207**:95 ff.

Elderly US population, drug consumption level, **207**:132

Electrochemical fluorination (ECF) PFOA, isomer profile (table), **208**:132

Electrochemistry, toxicity connection, **204**:137

Electron affinity, diacetyl relationship, **204**:138

Electron transfer & dermal toxicity, oxidative processes, **203**:120

Electron transfer & reactive oxygen species, dermatotoxic mechanisms, **203**:121

Electron transfer agents, chemical structures (illus.), **203**:121

Electron transfer groups, reduction potentials, **201**:42

Electron transfer groups, relation to bio-active agents, **201**:42

Electron transfer, dermal toxicity, **203**:119 ff.

Electron transfer, diacetyl toxicity, **204**:133 ff., 135

Electron transfer, oxidative stress, **204**:134

Electron transfer, physiological effects, **204**:134

Electron transfer, pulmonary toxicity mechanism, **201**:43

Electron transfer, pulmonary toxicity, **201**:41 ff.

Electron transfer, redox cycling & oxidative stress, **201**:42

Electron transfer, redox cycling, **204**:136

Electron transfer, RHOs, **206**:78

Electron transport effects, chromium (diag.), **210**:12

Electrostatics, action in living systems, **204**:138

Embryotoxicity in gammarids, pollutants, **205**:42

Emerging contaminants, San Francisco Bay, **206**:129

Emerging test species, *Gammarus* spp., **205**:64

Emulsification in water, oil spills, **206**:101

Endocarditis, *Pseudomonas aeruginosa*, **201**:77

Endocrine disruption in Gammarids, endpoints, **205**:45

Endocrine disruption, endpoints assessed, **209**:11

Endocrine disruption, metals, **210**:40

Endocrine disruption, overview, **202**:166

Endocrine system, aging effects, **207**:122

Endocytosis (phagocytosis), chemical transport, **203**:7

Endosulfan toxicity data, comparative distribution fit (illus.), **209**:47

Endosulfan, data-set distribution test (diag.),
   209:42
Endpoints, derivation methods context, **209**:11
Endrin toxicity data, comparative distribution
   fit (illus.), **209**:44
Endrin, data-set distribution test (diag.), **209**:41
Enterohepatic recirculation, PFOS, **202**:8
Environmental agents, anemia association
   (table), **207**:108
Environmental agents, hepatotoxicity (table),
   207:103–104
Environmental behavior, PFA structure effects,
   208:112
Environmental biomarker, HSPs in fish, **206**:17
Environmental biomonitors, fish, **206**:2
Environmental chemical sensitivity, aging
   humans, **207**:96
Environmental chemical skin penetration,
   aging effects, **207**:129
Environmental chemicals, elderly people,
   207:96
Environmental contaminants, illicit drug
   adulterants, **210**:81
Environmental contaminants, PFCs, **208**:179
Environmental contaminants, pharmaceuticals,
   207:19
Environmental contamination, chlorinated
   paraffins, **207**:78
Environmental contamination, chromium,
   210:2
Environmental contamination, clan labs,
   210:91
Environmental contamination, illicit drugs,
   210:59 ff.
Environmental contamination, PBDE, **207**:59
Environmental contamination, perfluorinated
   surfactants, **207**:51
Environmental contamination, pharmaceutical
   pathways (diag.), **202**:71
Environmental contamination, pharmaceuti-
   cals, **202**:74
Environmental contamination, pulmonary
   toxicity, **201**:41 ff.
Environmental degradation, fenoxycarb,
   202:168
Environmental detections, analgesics, **202**:98
Environmental detections, antibiotics, **202**:98
Environmental detections, antidepressants,
   202:101
Environmental detections, antiepileptics,
   202:101
Environmental detections, anti-inflammatories,
   202:98

Environmental detections, antineoplastic
   agents, **202**:101
Environmental detections,
   B_2-Sympathomimetics, **202**:102
Environmental detections, beta-blockers,
   202:99
Environmental detections, contrast media,
   202:102
Environmental detections, hormones &
   steroids, **202**:100
Environmental detections, lipid regulators,
   202:100
Environmental distribution, PCBs, **201**:139
Environmental drug contaminants, licit and
   illicit ones differ, **210**:71
Environmental endocrine disruption, overview,
   202:166
Environmental exposure studies, Red Dog
   Mine, **206**:57
Environmental exposure, chemicals (diag.),
   203:3
Environmental exposures, Great Lakes Region,
   207:1 ff.
Environmental exposures, sensitive
   populations, **207**:95 ff.
Environmental factors, bioavailability, **203**:27
Environmental factors, lipid-content effects,
   204:13
Environmental factors, PCB degradation
   effects (diag.), **201**:146
Environmental fate of PCBs, study methods,
   201:141
Environmental fate processes, PCBs (diag.),
   201:142
Environmental fate, coumaphos, **205**:150
Environmental fate, fenamiphos, **205**:134
Environmental fate, fenoxycarb, **202**:155 ff.
Environmental fate, human pharmaceuticals,
   202:110
Environmental fate, marine oil spills, **206**:100
Environmental fate, OP pesticides,
   205:117 ff., 121
Environmental fate, PCBs, **201**:137 ff.
Environmental fate, PFOS, **202**:5
Environmental fate, pharmaceuticals,
   202:53 ff.
Environmental fate, test regulations, **204**:2
Environmental impact, clan labs, **210**:91
Environmental impact, mining wastes, **206**:33
Environmental implications, ship oil spills,
   206:95 ff.
Environmental levels, chromium (table), **210**:4

Environmental loading effects, PFA sources, **208**:122

Environmental media contamination, by illicit drugs (table), **210**:77–79

Environmental media, chromium levels (table), **210**:4

Environmental occurrence, illicit drugs, **210**:76

Environmental oxidation, FTOH, **208**:166

Environmental partitioning, FTOHs, **208**:166

Environmental persistence, perfluorinated surfactants, **208**:162

Environmental PFA levels, precursors impact, **208**:82

Environmental pollutant, chromium, **210**:1

Environmental pollution, NTPs chemistry, **201**:117 ff.

Environmental pollution, stressed fish, **206**:1 ff.

Environmental release, pharmaceuticals, **202**:72

Environmental residues, perfluorocarboxylate isomers, **208**:148

Environmental residues, PFOA isomer content, **208**:121

Environmental residues, PFOA isomer profiles, **208**:137

Environmental residues, PFOA isomers (table), **208**:134

Environmental residues, PFOS isomers (table), **208**:144

Environmental sources, chromium, **210**:4

Environmental sources, illicit drugs, **210**:74

Environmental sources, morphine, **210**:74

Environmental stress, reactive oxygen species, **206**:1

Environmental stressor types, fish, **206**:3

Environmental stressors, HSPs, **206**:2

Environmental stressors, impact on fish, **206**:3

Environmental xenobiotics, immunotoxicity (table), **207**:121–122

Enzyme activator, diacetyl, **204**:145

Enzyme changes, induced by chromium, **210**:11

Enzyme inhibitors, glutathione-S-transferases, **204**:78

Enzyme role, pesticide metabolism, **204**:52

Enzyme-reaction scheme, cytochrome P450 (diag.), **204**:54

Enzymology, glutathione-S-transferases, **204**:77

Enzymology, pesticide metabolism (illus.), **204**:57

Epichlorohydrin, pulmonary toxin, **201**:50

Epidemiology studies, Red Dog Mine, **206**:57

Epidemiology, illicit drug contaminants, **210**:59 ff.

Epidemiology, role in FEUDS (Forensic Epidemiology Using Drugs in Sewage), **210**:83

Epidermal effects, aging, **207**:126

EqP theory (Equilibrium partitioning theory), invertebrates & chemicals, **203**:13

Equation symbols, bioavailability (table), **203**:65

Equilibrium partitioning theory (EqP theory), invertebrates & chemicals, **203**:13

Esterases, aquatic species pesticide metabolism, **204**:63

Esterases, substrates & inhibitors (illus.), **204**:57

Estimation techniques, filling chronic data gaps, **209**:103

Ethanol metabolites, diacetyl connection, **204**:140

Ethanol, pulmonary toxin, **201**:51

Ethoxylate contamination, Great Lakes surface waters & sediment, **207**:40

Ethylene oxide, pulmonary toxin, **201**:46

European Arctic, chemical contamination, **205**:78

European Arctic, organochlorine contamination, **205**:80

European Arctic, Svalbard glaucous gull bioindicator species, **205**:77 ff.

Evaporation loss, oil spills, **206**:101

Even fluorotelomer alcohols (FTOHs), atmospheric lifetime, **208**:48

Ex situ site remediation, chromium, **210**:17

Exceedance limits, frequency, **209**:71

Excretion rate, gammarids, **205**:32

Excretion rates, pharmaceuticals (table), **202**:70

Exposure modes, assessing gammarids (table), **205**:54

Exposure reduction, home visits & surveys, **201**:24

Exposure sources, formaldehyde, **203**:106

Exposure to house dust, infants, **201**:1 ff.

Exposure to house dust, research needed, **201**:29

Exposure types, gammarids, **205**:53

Exposure, to banned pesticides, **201**:8

Exposures of sensitive populations, the elderly, **207**:95 ff.

Extraction methods, organic pollutants in soil,
  205:131
Eye infections, *Pseudomonas aeruginosa*,
  201:82

F
Facilitated transport, chemicals, **203**:6
Farm worker exposure, pesticides, **201**:7
Fate and occurrence, human pharmaceuticals,
  202:53 ff.
Fate in animals, isofenphos, **205**:149
Fate in drinking water, pharmaceuticals,
  202:134
Fate in environment, fenoxycarb, **202**:155 ff.
Fate in environment, marine oil spills, **206**:100
Fate in environment, pharmaceuticals, **202**:110
Fate in humans, isofenphos, **205**:149
Fate in mammals, fenamiphos, **205**:142
Fate in soil, isofenphos, **205**:154
Fate in soil, pharmaceuticals, **202**:129, 132
Fate in water, coumaphos, **205**:151
Fate in water, isofenphos, **205**:145
Fate of oil spills, marine environment (diag.),
  206:101
Fate of oils spills, weathering, **206**:100
Fate, illicit drugs, **210**:94
Fate, perfluoroacyl fluorides, **208**:4
Fate, perfluoroacyl radicals (table), **208**:44
FAV (final acute value) (table), PFOS, **202**:27
Feeding activity test methods, gammarids
  (table), **205**:11
Feeding activity testing, gammarids, **205**:10
Feeding rate and uptake, gammarid toxicity,
  205:19
Fenamiphos acute toxicity, animals (table),
  205:143
Fenamiphos acute toxicity, aquatic and
  terrestrial species (table), **205**:141
Fenamiphos and metabolites, leaching
  behavior, **205**:137
Fenamiphos and metabolites, soil sorption
  (tables), **205**:136
Fenamiphos, chronic toxicity, **205**:143
Fenamiphos, description, **205**:133
Fenamiphos, ecotoxicology, **205**:139
Fenamiphos, environmental fate, **205**:134
Fenamiphos, fate and toxicity, **205**:117 ff.
Fenamiphos, mammalian fate and toxicity,
  205:142
Fenamiphos, microbial effects, **205**:137
Fenamiphos, physiochemical properties
  (table), **205**:134
Fenamiphos, plant metabolism, **205**:138

Fenamiphos, soil behavior, **205**:135
Fenamiphos, water residues (table), **205**:135
Fenamiphos, wildlife effects, **205**:141
Fenoxycarb & metabolites, structures (diag.),
  202:161
Fenoxycarb, aerobic aquatic metabolism
  (table), **202**:177
Fenoxycarb, chemistry & fate, **202**:155 ff.
Fenoxycarb, chemistry, **202**:159
Fenoxycarb, description, **202**:155
Fenoxycarb, environmental degradation,
  202:168
Fenoxycarb, field dissipation, **202**:171
Fenoxycarb, hydrolysis, **202**:168
Fenoxycarb, insect growth regulator (IGR),
  202:155
Fenoxycarb, photolytic pathway (diag.),
  202:170
Fenoxycarb, physicochemical properties
  (table), **202**:160
Fenoxycarb, soil adsorption (table), **202**:162
Fenoxycarb, soil metabolism (diag.), **202**:175
Fenoxycarb, soil persistence (table), **202**:178
Fenoxycarb, soil persistence, **202**:171
Fenoxycarb, synthesis pathway (diag.),
  202:164
Fertilizer contamination, chromium, **210**:5
FEUDS (Forensic Epidemiology Using Drugs
  in Sewage) & illicit drugs, future utility,
  210:96
FEUDS illicit drug source reconstruction,
  210:83
FEUDS vs. population surveys, advantages,
  210:85
FEUDS, chronological publication listing
  (table), **210**:87
FEUDS, legal concerns, **210**:89
FEUDS, methods and limitations, **210**:84
FEUDS, quality assurance, **210**:86
Fick's Law, chemical uptake, **203**:7
Field data, in criteria derivation, **209**:13
Field dissipation, fexoxycarb, **202**:171
Final acute value (FAV) (table), PFOS, **202**:27
Fish & pollution, heat shock proteins, **206**:1 ff.
Fish advisory, PFOS residues, **208**:194
Fish bioconcentration, PFOS, **202**:10
Fish chronic toxicity, PFOS, **202**:21
Fish contaminant, PFOS, **208**:188
Fish contamination, by long-chain
  perfluorinated substances (table),
  208:187
Fish contamination, PFCs (table), **208**:181
Fish critical body burdens, PROS, **202**:28

Fish effects, spilled petroleum, **206**:105

Fish genes, HSPs, **206**:11

Fish HSP expression, seasonal influence, **206**:15

Fish HSP overexpression, stress defense, **206**:17

Fish residues, flame retardant congener distribution (illus.), **207**:68

Fish residues, flame retardants, **207**:70

Fish residues, PFCs, **208**:180

Fish residues, PFOS & PFOA, **208**:180

Fish residues, surfactants, **207**:40

Fish residues, synthetic musks, **207**:48

Fish response, stress, **206**:5

Fish toxicity, fenamiphos and metabolites (table), **205**:140

Fish, antioxidant biomarkers of pollution, **206**:5

Fish, as environmental biomonitors, **206**:2

Fish, BCF for PFOS (table), **202**:9

Fish, bioconcentration and PFC carbon number (diag.), **202**:40

Fish, chemical stressors, **206**:3

Fish, developmental role of HSP, **206**:10

Fish, environmental stressors, **206**:3

Fish, impact of environmental stressors, **206**:3

Fish, metal-induced stress, **206**:4

Fish, oil spill effects, **206**:104

Fish, PFOS toxicity, **202**:15

Fish, role of stress proteins, **206**:7

Fish, stress hormone release, **206**:6

Flame retardant congeners, fish residues (illus.), **207**:68

Flame retardant contaminants, Great Lakes, **207**:70

Flame retardant residues, Great Lakes basin, **207**:74

Flame retardant residues, Great Lakes fish, **207**:70

Flame retardant residues, Great Lakes samples (table), **207**:75–76

Flame retardant residues, Great Lakes, **207**:64

Flame retardant residues, gulls eggs (illus.), **207**:73

Florey-Higgens theory, chemical interactions, **204**:8

Fluorinated alkyl substances, biodegradation, **208**:161 ff.

Fluorinated alkyl substances, names & structures (table), **208**:163

Fluorinated anesthetics, atmospheric sources & levels, **208**:61

Fluorinated chemical toxicity, aquatic plants (diag.), **202**:45

Fluorinated organics, biodegradability, **208**:171

Fluorinated polymers, biodegradation, **208**:170

Fluorinated polymers, description & role, **208**:170

Fluorotelomer acrylates (FTAcs), atmospheric lifetime, **208**:54

Fluorotelomer alcohols (FTOHs), structure (diag.), **202**:2

Fluorotelomer carboxylic acids (FTCA), aquatic species toxicity (diag.), **202**:43

Fluorotelomer compounds, atmospheric sources, **208**:66

Fluorotelomer ethoxylates, degradation behavior, **208**:168

Fluorotelomer iodides (FTIs), atmospheric degradates, **208**:53

Fluorotelomer iodides, atmospheric lifetime, **208**:53

Fluorotelomer olefins (FTOs), atmospheric lifetime, **208**:51

Fluorotelomer-based chemicals, biodegradation, **208**:165

Fluorotelomers, carbon atom kinetics (table), **208**:37

Fluorotelomers, general information, **208**:35

Fluorotelomers, hydroxyl radical kinetics (table), **208**:39

Flurotelomer aldehydes (FTALs), photolysis properties (table), **208**:43

Food choice experiments, *Gammarus*, **205**:16

Food coating uses, PFCs, **208**:195

Food contamination, PFCs, **208**:180, 195

Food supply contamination, illicit drugs, **210**:81

Forensic Epidemiology Using Drugs in Sewage (FEUDS), illicit drug source assessment, **210**:83

Forensic epidemiology, illicit drugs, **210**:59 ff.

Formaldehyde toxicity, nervous system, **203**:105 ff.

Formaldehyde, absorption & distribution, **203**:107

Formaldehyde, behavioral effects, **203**:111

Formaldehyde, chronic effects, **203**:108

Formaldehyde, exposure sources, **203**:106

Formaldehyde, macro-molecular binding, **203**:111

Formaldehyde, metabolic behavior, **203**:106

Formaldehyde, neuronal morphology effects, **203**:110

Formaldehyde, neuro-oxidative effects,
    203:109
Formaldehyde, neurotoxicity, 203:108
Formaldehyde, physico-chemical features,
    203:105
Formaldehyde, pulmonary toxin, 201:47
Formaldehyde, sources & uses, 203:106
Formaldehyde, toxic effects, 203:107
Fragrance residues, surface waters (table),
    207:31
Frequency of exceedance, limits, 209:71
Frequent detects, drugs of abuse (table),
    210:75
Freshwater diatoms, pesticide effects,
    203:87 ff.
Freshwater invertebrates and fish, chemical
    toxicity (table), 205:9
Freundlich isotherm, equation, 203:15
Frustule (diatom cell wall), chemical effects,
    203:90
FTAcs (fluorotelomer acrylates), atmospheric
    degradates, 208:55
FTAcs, atmospheric concentrations (table),
    208:78
FTAcs, atmospheric levels, 208:77
FTALs (flurotelomer aldehydes), atmospheric
    degradates, 208:46
FTALs, atmospheric lifetime, 208:45
FTCA, aquatic species toxicity (diag.),
    202:43
FTIs (fluorotelomer iodides), atmospheric
    degradates, 208:53
FTOH (even fluorotelomer alcohol) behavior,
    environmental partitioning, 208:166
FTOH, environmental oxidation, 208:166
FTOH, soil biodegradation pathway (diag.),
    208:167
FTOHs (fluorotelomer alcohols, structure
    (diag.), 202:2
FTOHs, atmospheric concentrations (table),
    208:68
FTOHs, atmospheric degradates, 208:49
FTOHs, atmospheric levels, 208:67
FTOHs, biodegradation, 208:165
FTOs (fluorotelomer olefins), atmospheric
    concentrations (table), 208:76
FTOs, atmospheric levels, 208:67
Fugitive dust exposure, Red Dog Mine,
    206:49 ff., 50
Fullerenes, human skin effects, 203:129
Fungal bioremediation, chromium, 210:15
Future changes San Francisco Bay, water
    quality, 206:138

G
G. locusta, coastal sediment toxicity, 205:8
G. pulex, contaminated sediment toxicity,
    205:8
Gametogenesis activity, Gammarus spp.,
    205:48
Gammarid biomarker, chitin and molting,
    205:46
Gammarid biomarkers, heat shock proteins,
    205:45, 48
Gammarid effects, pulsed exposure models,
    205:57
Gammarid exposure assessment, modes
    (tables), 205:54
Gammarid exposure, types, 205:53
Gammarid feeding activity and rate, modeling,
    205:19
Gammarid feeding activity, toxicity effects,
    205:17
Gammarid feeding ecology, parasite effects,
    205:21
Gammarid parasites, antipredator effects,
    205:26
Gammarid reproduction, pesticide effects,
    205:44
Gammarid sensitivity, metal toxicity, 205:8
Gammarid test methods, evaluating existing
    ones, 205:62
Gammarid time-response assays, testing
    method, 205:10
Gammarid toxicity testing, feeding activity
    (table), 205:11
Gammarid toxicity, behavioral test methods
    (table), 205:23
Gammarid toxicity, feeding rate, uptake and
    depuration, 205:19
Gammarid toxicity, PAHs (polyaromatic
    hydrocarbons), 205:8
Gammaridea, natural habitat, 205:2
Gammarids toxicity, population dynamics
    effect, 205:43
Gammarids, behavioral assays, 205:22
Gammarids, bioenergetic responses, 205:32
Gammarids, endocrine disruption endpoints,
    205:45
Gammarids, excretion and respiration rates,
    205:32
Gammarids, feeding activity test methods
    (table), 205:11
Gammarids, feeding activity testing, 205:10
Gammarids, in situ testing approaches, 205:58
Gammarids, in situ vs. ex situ test results,
    205:18

Gammarids, laboratory vs. field tests, **205**:18

Gammarids, lethality testing, **205**:7

Gammarids, metallothionein induction, effects, **205**:42

Gammarids, mode-of-action studies, **205**:32

Gammarids, mode-of-action test methods (table), **205**:33

Gammarids, predation effects, **205**:20

Gammarids, sediment toxicity assays, **205**:57

Gammarids, use in biomarker tests, **205**:32

Gammarus spp., aquatic ecotoxicology, **205**:1 ff.

Gammarus spp., biomarkers (table), **205**:38

Gammarus spp., food choice experiments, **205**:16

Gammarus spp., gametogenesis activity, **205**:48

Gammarus spp., multimetric test systems, **205**:64

Gammarus, as indicator species, **205**:5

Gammarus, behavior and character, **205**:2

Gammarus, behavior testing, **205**:4

Gammarus, culturing methods, **205**:5

Gammarus, drift behavior and pollution, **205**:22

Gammarus, drift behavior and predators, **205**:22

Gammarus, emerging test species, **205**:5

Gammarus, feeding behavior, **205**:3

Gammarus, in water quality testing, **205**:1 ff.

Gammarus, leaf-mass feeding assays, **205**:17

Gammarus, life cycle, **205**:2

Gammarus, mating behavior, **205**:3

Gammarus, species distribution, **205**:2

Gammarus, toxicity endpoints, **205**:4

Gaseous pollutant removal, NTPs chemistry, **201**:121

Gastrointestinal infections, *Pseudomonas aeruginosa*, **201**:79

General public exposure, designer drugs, **210**:82

Generalized stress response, fish, **206**:5

Genetic engineering, phytoremediation improvement, **210**:23

Genotoxicity bioassay, bioavailability, **203**:33

Genotoxicity of contaminants, Svalbard glaucous galls, **205**:100

Genotoxicity, diacetyl, **204**:144

Genotoxicity, diatom nucleus, **203**:89

Genus mean acute toxicity values (GMAV), PFOS (table), **202**:26

Genus mean chronic values (GMCV), PFBS (table), **202**:30

Geography, Red Dog Mine (illus.), **206**:50

Geraniol, toxicity, **203**:128

Germination in plants, chromium inhibition, **210**:9

Glaucous gulls, legacy chiral organochlorines, **205**:81

Glaucous gulls, organohalogen contaminates (illus.), **205**:80

GLI (Great Lakes Initiative), water quality criteria, **202**:22

Global distribution, PCBs, **201**:137 ff., 139

Glomerular filtration, aging effects, **207**:98

Glucocorticoid effects, Svalbard glaucous gall residues, **205**:96

Glucose transferases, pesticide metabolism, **204**:72

Glucuronic acid transferases, pesticide metabolism, **204**:72

Glutathion, HSF (heat shock factor) relationship, **206**:13

Glutathione-S-transferase activity, detoxification biomarker, **205**:52

Glutathione-S-transferase, substrates (illus.), **204**:57

Glutathione-S-transferases, aquatic organisms, **204**:75

Glutathione-S-transferases, enzymology, **204**:77

Glutathione-S-transferases, inhibitors, **204**:78

GMAV (genus mean acute toxicity values), PFOS (table), **202**:26

GMAV and GMCV (genus mean chronic values) (table), PFBS, **202**:30

GMAV, PFOA (table), **202**:32

GMCF (genus mean chronic values), PFBS (table), **202**:30

GMCV, PFOA (table), **202**:32

Goals, UCDM, **209**:95

Gonadal steroid hormones, Svalbard glaucous gall effects, **205**:95

Gram-negative vs. –positive bacteria, characteristics, **203**:9

Granulocytopenic patients, *Pseudomonas aeruginosa*, **201**:86

Great Lakes Basin, chemical contamination, **207**:1 ff.

Great Lakes basin, flame retardant residues, **207**:74

Great Lakes biota & sediment, PBDE sampling locations (illus.), **207**:60

Great Lakes bird contamination, PBDE, **207**:63

Great Lakes chemical contaminants, statistical treatment, **207**:4

Great Lakes contamination, environmental
   exposure analysis, **207**:1 ff.
Great Lakes ecosystem, government
   agreements, **207**:2
Great Lakes fish, organochlorine residues,
   207:62
Great Lakes fish, PBDE residues, **207**:61
Great Lakes Initiative (GLI), water quality
   criteria, **202**:22
Great Lakes pollutants, current-use pesticides,
   207:6
Great Lakes pollution, pesticide detection,
   207:14
Great Lakes pollution, pesticides, **207**:12
Great Lakes residues, alkyphenol ethoxylate
   (table), **207**:42–43
Great Lakes residues, bisphenol A, **207**:33,
   207:34
Great Lakes residues, chlorinated paraffins
   (table), **207**: 81
Great Lakes residues, industrial chemicals,
   207:34
Great Lakes water & sediment pollution,
   surfactants, **207**:38
Great Lakes waters, perfluorinated surfactant
   contaminants, **207**:52
Great Lakes waters, DEET & phthalate
   detections, **207**:36
Great Lakes waters, musk residues,
   207:49
Great Lakes waters, personal care products,
   207:35
Great Lakes waters, triclosan detections,
   207:35
Great Lakes Watershed, contaminant threats,
   207:3
Great Lakes watershed, perfluorinated
   compound contamination, **207**:55
Great Lakes, current-use pesticide levels
   (illus.), **207**:18
Great Lakes, flame retardant contamination,
   207:70
Great Lakes, pesticide sampling locations
   (illus.), **207**:13
Great Lakes, pharmaceutical sampling
   locations (illus.), **207**:23
Great Lakes, synthetic musk residues, **207**:47
Ground water detections, pharmaceuticals
   (table), **202**:85–89
Groundwater residues, fenamiphos (table),
   205:135
Gull eggs, flame retardant residues (illus.),
   207:73

H
Half-lives in soil, OP pesticides (table),
   205:130
Hand washing, contamination and disease
   prevention, **201**:23
Harmonization for sediments, criteria setting,
   209:92
Hazards to humans and wildlife, OP pesticides,
   205:118
HBCD (hexabromocyclodecane) residues,
   biota & sediment, **207**:70
HCFC (hydrochlorofluorocarbon), carbon
   atom kinetics (table), **208**:18
HCFC, hydroxyl radical kinetics (table),
   208:18
HCFC-123, reaction kinetics, **208**:20
HCFC-124, tropospheric fate, **208**:16
HCFCs atmospheric fate, **208**:16
HCFCs, atmospheric degradates, **208**:17, 21
HCFCs, atmospheric lifetimes, **208**:21
HCFCs, atmospheric sources & levels, **208**:61
Health cost reduction, home visits, **201**:25
Health effects, OP pesticides, **205**:132
Health guidance values, are they adequate?,
   207:134
Healthy-human infections, *Pseudomonas
   aeruginosa* (table), **201**:75
Hearing effects, aging, **207**:113
Heat shock factors (HSF), heat shock response,
   206:12
Heat shock proteins (HSP), stressed
   fish, **206**:1 ff.
Heat shock proteins, gammarid biomarkers,
   205:45, 48
Heat stress proteins, stress response, **206**:2
Heat-stress response, muted in Antarctic fish,
   206:10
Heavy metal content, medicinal plants (table),
   203:141
Heavy metal effects, medicinal plants,
   203:139 ff.
Heavy metal effects, plant cell signaling
   (diag.), **203**:144
Heavy metal effects, plant growth
   & metabolism, **203**:140
Heavy metal exposure, plant effects, **210**:13
Heavy metal toxicity, plants, **203**:140
Heavy metals, mechanism of action, **203**:142
Heavy metals, plant effects, **203**:139
Heavy metals, plant response strategies (illus.),
   210:13
Heavy metals, toxicity & exposure, **206**:51

Heavy-metal human exposure, Red Dog Mine, **206**:49 ff.

Heavy-metal stressed plants, metabolite production (table), **203**:143

Hematopoiesis, aging effects, **207**:106

Hepatic cytochromes, aging effects, **207**:102

Hepatic function, changes with aging, **207**:100

Hepatotoxicity, causative agents (table), **207**:103–104

Heptachlor toxicity data, comparative distribution fit (illus.), **209**:46

Heptachlor, data-set distribution test (diag.), **209**:42

Herbicide detections, surface waters, **207**:19

Herbicide effects in algae, growth, **203**:93

Herbicide effects, algal communities, **203**:96

Herbicide effects, algal nutrient absorption, **203**:93

Herbicide effects, diatom cell density, **203**:94

Herbicide effects, diatom cytoskeleton, **203**:89

Herbicide effects, diatom species, **203**:94

Herbicide inhibition, algal lipids & carbohydrates, **203**:92

Herbicides, algal biosynthesis effects, **203**:92

Herbicides, algal photosynthesis inhibition, **203**:91

Herbicides, deriving chronic criteria, **209**:117

Heroin, biodegradation, **210**:80

HFC-125 & -134a, atmospheric levels, **208**:62

HFC-125, atmospheric lifetime, **208**:27

HFC-134a, atmospheric concentrations (table), **208**:63

HFCs (hydrofluorocarbons), atmospheric sources & levels, **208**:62

HFCs, atmospheric degradates, **208**:26

HFCs, atmospheric lifetime of saturated forms, **208**:22

HFCs, chlorine atom kinetics (table), **208**:23

HFCs, hydroxyl radical kinetics (table), **208**:23

HFCs, stratospheric transport, **208**:26

HFOs (hydrofluoroolefins), atmospheric degradates, **208**:30

HFOs, atmospheric lifetime, **208**:30

HFOs, chlorine atom kinetics (table), **208**:31

HFOs, hydroxyl radical kinetics (table), **208**:31

HFOs, uses, **208**:66

Hill reaction inhibition in algae, herbicides, **203**:92

Home cleaning, removing pollutants, **201**:19

Home pollutant problems, discussion, **201**:28

Home surveys, toxicant exposure reduction, **201**:24

Home visits, toxicant exposure reduction, **201**:24

Hormone detections, surface water & sediments, **207**:36

Hormone effects, Svalbard glaucous gall residues, **205**:94

Hormone residues, surface waters (table), **207**:31–32

Hormones & steroids, environmental detections, **202**:100

Hormones and steroids, description, **202**:66

Hormones, organic wastewater contamination, **207**:30

Hospital admissions, drug interactions, **207**:133

Hot water pollutant extraction, carpet vacuuming, **201**:22

House dust and allergens, infants and children, **201**:2

House dust and exposure, research needed, **201**:29

House dust contaminant, phthalates, **201**:13

House dust contaminants, pesticides & metals, **201**:3

House dust contamination, PFCs (table), **208**:205

House dust contamination, PFOS & PFOA (table), **208**:205

House dust contamination, various chemicals (table), **201**:9, 15

House dust pollutants, exposure, **201**:2

House dust residues, PAHs (table), **201**:12

House dust residues, perfluorinated compounds, **208**:204

House dust, allergens & dust mites, **201**:18

House dust, Bacteria, viruses, mold, **201**:18

House dust, EDCs, **201**:14, 17

House dust, home cleaning, **201**:19

House dust, monitoring & sampling methods, **201**:4

House dust, monitoring and exposure, **201**:1 ff.

House dust, monitoring pollutants, **201**:3

House dust, mutagens, **201**:3

House dust, pesticides, **201**:6

House dust, three-spot vacuum test, **201**:21

House dust, toxicants, **201**:6

House residues, PFOS & PFOA, **208**:204

HSF link, thiol-containing molecules, **206**:13

HSF, glutathion relationship, **206**:13

HSP (heat shock protein), description, **206**:2

HSP expression, seasonal influences, **206**:15

HSP families, molecular chaperones, **206**:6

HSP genes, fish, **206**:11

HSP induction, mechanistic regulation, **206**:12

HSP overexpression in fish, stress defense, **206**:17

HSP production, chemical pollutants, **206**:7

HSP production, role in animal adaptation, **206**:8

HSP, apoptosis effect, **206**:14

HSP, cross tolerance in fish, **206**:11

HSP, developmental role in fish, **206**:10

HSP, pollutant protective response, **206**:10

HSP, protein metabolism interaction, **206**:6

HSP, relationship to P450 inducers, **206**:7

HSP, role in survival, **206**:14

HSP, stress-induced production, **206**:6

Human adverse effects, lead exposure (diag.), **206**:52

Human aging, environmental sensitivity, **207**:96

Human behavior effects, metal contamination, **210**:42

Human behavioral response, caesium, **210**:43

Human behavioral response, non-radioactive Cd & Co, **210**:43

Human behavioral response, rodent model, **210**:44

Human behavioral response, uranium, **210**:42

Human carcinogen, chromium, **210**:2

Human chemical exposure, soil ingestion

Human consequences, radionuclide effects, **210**:49

Human distribution, PFOA isomers (table), **208**:128

Human exposure to metals, Red Dog Mine, **206**:49 ff.

Human exposure, contaminated soil contact, **206**:54

Human exposure, perfluorinated substances, **208**:179 ff.

Human exposure, zinc & cadmium, **206**:51

Human food, perfluorinated substances, **208**:179 ff.

Human health data, role in UCDM, **209**:101

Human health effects, chromium, **210**:8

Human health effects, *Pseudomonas aeruginosa* (table), **201**:75

Human pathogenicity, *Pseudomonas* spp., **201**:74

Human pharmaceuticals, occurrence and fate, **202**:53 ff.

Human residues, PFOS isomers (table), **208**:140

Humans, known renal toxicants (table), **207**:99–100

Humic acids, bioconcentration effects, **204**:18

HVS3 (high-volume low-surface), pollutant sampling, **201**:4

Hydrocarbons from petroleum, water solubility, **206**:99

Hydrochlorofluorocarbons (HCFCs), atmospheric fate, **208**:16

Hydrofluorocarbons (HFCs), atmospheric sources, **208**:62

Hydrofluoroolefin (HFO) mechanism, perfluoroacyl halide formation, **208**:7

Hydrogen cyanide, pulmonary toxin, **201**:47

Hydrogen sulfide, pulmonary toxin, **201**:48

Hydrolysis affects, OP pesticides, **205**:126

Hydrolysis half lives, OP pesticides (table), **205**:124

Hydrolysis lifetimes, perfluoroacyl halides, **208**:4

Hydrolysis, fenoxycarb, **202**:168

Hydrolysis, perfluoroacyl fluorides, **208**:4

Hydrolysis, PFOS, **202**:6

Hydrophobic chemical uptake, microbes, **203**:24

Hydrophobic chemicals, organismal uptake, **203**:8

Hydrophobicity, BCF effects, **204**:8

Hydroxyl radical kinetics, volatile anesthetics (table), **208**:14

Hygiene hypothesis, children's environmental exposure, **201**:19

Hyperaccumulator plants, for phytoremediation, **206**:35

Hyperaccumulator plants, listing (table), **206**:39

Hyperaccumulator plants, response types, **210**:13

Hyperaccumulators, plant species, **206**:42

Hypochlorous acid, pulmonary toxin, **201**:48

Hypothesis testing issues, ecotoxicity, **209**:9

Hypothesis testing, ecotoxicity evaluation, **209**:9

I

IGR (insect growth regulator), fenoxycarb, **202**:155

Illegal oil discharges, Baltic Sea (illus.), **206**:109

Illegal use, livestock & racing animal drugs, **210**:92

Illicit drug adulterants, examples, **210**:81

Illicit drug adulterants, potential environmental contaminants, **210**:81

Illicit drug adulterants, role, **210**:81

Illicit drug environmental contamination, origins, **210**:60

Illicit drug impurities, potential environmental contaminants, **210**:81

Illicit drug source assessment, FEUDS, **210**:83

Illicit drug source assessment, sewage epidemiology (forensics), **210**:83

Illicit drug source reconstruction, FEUDS, **210**:83

Illicit drug source, leftover medication, **210**:94

Illicit drug, definition, **210**:67

Illicit drugs & FEUDS, future utility, **210**:96

Illicit drugs in the environment, publication chronology (table), **210**:63–66

Illicit drugs in the environment, sources & routes (diag.), **210**:62

Illicit drugs, adulterants & impurities (table), **210**:69–70

Illicit drugs, categorization, **210**:73

Illicit drugs, counterfeiting, **210**:68

Illicit drugs, dermal contact & transfer, **210**:93

Illicit drugs, description & terminology, **210**:67

Illicit drugs, detected in environmental media (table), **210**:77–79

Illicit drugs, diversion from legitimate use, **210**:93

Illicit drugs, ecotoxicity, **210**:94

Illicit drugs, environmental contaminants, **210**:59 ff.

Illicit drugs, environmental occurrence, **210**:76

Illicit drugs, environmental sources, **210**:74

Illicit drugs, fate and transport, **210**:94

Illicit drugs, food supply contamination, **210**:81

Illicit drugs, in ambient air, **210**:91

Illicit drugs, in the money supply, **210**:90

Illicit drugs, includes legal pharmaceuticals, **210**:68

Illicit drugs, source assessment, **210**:82

Illicit vs. licit drugs as contaminants, differences, **210**:71

Illicit vs. licit drugs, characteristics, **210**:70

Illness in spas, *Pseudomonas aeruginosa* (table), **201**:98

Immune responsiveness, aging effects, **207**:116

Immune system, aging effects, **207**:116

Immunity, Svalbard glaucous galls, **205**:98

Immunocompromised humans, *Pseudomonas aeruginosa* (table), **201**:76

Immunotoxicity, from xenobiotics (table), **207**:121–2

Impurities, in illicit drugs (table), **210**:69–70

In situ site remediation, chromium, **210**:17

In situ testing approaches, gammarids, **205**:58

In situ vs. *ex situ* test results, gammarids, **205**:18

India mining slimes, iron content (table), **206**:33

India, iron-ore beneficiation process, **206**:32

Indirect food contamination, PFCs, **208**:195

Indoor air pollution, house dust, **201**:3

Indoor air purification, NTPs chemistry, **201**:124

Inducers of P450, HSP implications, **206**:7

Inducers, oxidases (illus.), **204**:57

Industrial chemical residues, Great Lakes, **207**:34

Infant exposure, house dust, **201**:1 ff.

Infant hand washing, preventing contamination, **201**:23

Infants, relative toxicant exposure, **201**:2

Infectious diseases, hand washing, **201**:23

Infective dose, *Pseudomonas aeruginosa*, **201**:102

Inhalation of fugitive dust, Red Dog Mine, **206**:53

Inhibitors, esterases (illus.), **204**:57

Inhibitors, oxidases (illus.), **204**:57

Innate immunity effects, aging, **207**:117

Insect repellant residues, surface waters (table), **207**:31

Insecticide detections, surface waters, **207**:19

Insecticide metabolism in aquatic species, organophosphates (table), **204**:64

Insecticides, pulmonary toxins, **201**:56

Integumentary system, aging effects, **207**:126

International incidence, oil spills, **206**:96

International Joint Commission (IJC), chemicals of emerging concern, **207**:2

International regulation, environmental pollutants (table), **207**:7–11

Interspecies correlations, QSAR use, **209**:16

Invertebrate contamination, PFCs (table), **208**:189

Invertebrate toxicity, PFOS, **202**:14

Invertebrates, cadmium behavioral effects, **210**:47

Invertebrates, chemical uptake, **203**:12

Invertebrates, PFC levels, **208**:193

Iodide uptake effects, thyroid follicular cells (table), **207**:123

Iron content, Iron-ore slimes from mining (table), **206**:33

Iron-ore beneficiation, India, **206**:32

Iron-ore mining wastes, phytoremediation treatment, **206**:31

Iron-ore mining, metal toxicity, **206**:31
Iron-ore slimes, composition (table), **206**:33
Iron-ore tailings phytoaccumulation, lemon
    grass (illus.), **206**:35
Iron-ore tailings treatment, tomato plants
    (illus.), **206**:36
Iron-ore tailings treatment, tree species (illus.),
    206:36
Iron-ore tailings utilization, phytoremediation,
    206:34
Iron-ore tailings, characterization, **206**:32
Iron-ore tailings, composition, **206**:32
Iron-ore tailings, contaminants, **206**:32
Iron-ore tailings, nature and production,
    206:31
Iron-ore wastes, from mining, **206**:30
Iron-ore wastes, phytoremediation, **206**:29 ff.
Isofenphos, aquatic species effects, **205**:146
Isofenphos, carcinogenicity, **205**:148
Isofenphos, chronic toxicity, **205**:147
Isofenphos, description, **205**:144
Isofenphos, effects on birds, **205**:146
Isofenphos, fate in humans and animals,
    205:149
Isofenphos, fate in soil, **205**:145
Isofenphos, fate in water, **205**:145
Isofenphos, mammalian toxicity, **205**:147
Isofenphos, mutagenicity, **205**:148
Isofenphos, reproductive toxicity, **205**:148
Isofenphos, teratogenic effects, **205**:148
Isoflurane lifetime, atmospheric fate, **208**:15
Isomer composition, PFOA & PFOS (table),
    208:116
Isomer composition, PFOA (table), **208**:132
Isomer nomenclature, PFAs, **208**:112
Isomer profiles, PFOA & PFOS (illus.),
    208:120
Isomer profiles, PFOS fluoride-derived
    products (table), **208**:121
Isomer profiling, PFA substances, **208**:111 ff.
Isomer separation methods, PFAs, **208**:125
Isomer separation, PFOSs (illus.), **208**:130
Isomerism, bioconcentration effects, **204**:9
Isomer-specific methods, perfluoroalkyl
    compounds, **208**:123
Issues for San Francisco Bay, South Bay Salt
    Pond Restoration, **206**:115 ff.

J
JHA (juvenile hormone agonist), defined,
    202:155
JHA, defined, **202**:155
JHAs, mode of action, **202**:164

JHAs, names & structures (diag.), **202**:158
Juvenile hormones, structures (diag.), **202**:157

K
Kidney, implications of human aging, **207**:97
Kinetics of fluorotelomers, hydroxyl
    radical-initiated (table), **208**:39
Kinetics of flurotelomers, carbon atom-initiated
    (table), **208**:37
Kinetics of HCFCs, carbon atom-initiated
    (table), **208**:18
Kinetics of HCFCs, hydroxyl radical-initiated
    (table), **208**:18
Kinetics of HFCs, chlorine atom-initiated
    (table), **208**:23
Kinetics of HFCs, hydroxyl radical-initiated
    (table), **208**:23
Kinetics of HFOs, chlorine atom-initiated
    (table), **208**:31
Kinetics of HFOs, hydroxyl radical-initiated
    (table), **208**:31
Kinetics of perfluorosulfonamides, carbon
    atom-initiated, (table), **208**:57
Kinetics of perfluorosulfonamides, hydroxyl
    radical-initiated (table), **208**:57
Kinetics, uptake by organisms, **203**:7
Kweon classification scheme, RHOs (table),
    206:71–72
Kweon classification scheme, RHOs, **206**:70

L
Laboratory vs. field tests, gammarids, **205**:18
Lawn care chemicals, Great Lakes pollution,
    207:14
LC-MS/MS analysis, PFOS isomers (illus.),
    208:130
Leaching behavior, fenamiphos, **205**:137
Leaching potential, OP pesticides (table),
    205:124
Leaching potential, OP pesticides, **205**:125
Lead exposure, children, **206**:56
Lead exposure, Red Dog Mine workers, **206**:58
Lead exposure, Red Dog Mine, **206**:52
Lead exposure, subsistence land use, **206**:56
Lead exposure, wildlife effects, **206**:56
Lead health effects, humans (diag.), **206**:52
Lead in dust, child lead blood levels, **201**:6
Lead levels, Red Dog Mine workers (illus.),
    206:59
Lead measurement, carpets, **201**:4
Lead, in dust-bearing soils (illus.), **206**:55
Lead-zinc mine, Alaska, **206**:49
Leaf-mass feeding assays, *Gammarus*, **205**:17
Legal concerns, FEUDS, **210**:89

Legal pharmaceuticals, and illicit drugs, 210:68

Lemon grass, phytoremediation (illus.), 206:35

Lethality testing, gammarids, 205:7

Leukemia, organochlorines in house dust, 201:11

Licit vs. illicit drugs as contaminants, differences, 210:71

Licit vs. illicit drugs, characteristics, 210:70

Life cycle, *Gammarus*, 205:2

Light effects, bioconcentration, 204:16

Limitations to derived criteria, review, 209:133

Limitations, UCDM, 209:93

Linalool, toxicity, 203:129

Lindane toxicity data, comparative distribution fit (illus.), 209:45

Lindane, data-set distribution test (diag.), 209:41

Lipid content in aquatic species, bioconcentration effects, 204:9

Lipid content, aquatic species (table), 204:10

Lipid peroxidation, biomarker for metal exposure, 205:49

Lipid regulators, description, 202:67

Lipid regulators, environmental detections, 202:100

Lipid-content effects, environmental factors, 204:13

Liver, aging effects, 207:100

Livestock drugs, illegal use, 210:92

Log-normal distribution fit, pesticide data sets (table), 209:51

Log-triangular distribution fit, pesticide data sets (table), 209:51

Long-chain perfluorinated substances, edible fish residues (table), 208:187

Long-term toxicity, fenamiphos, 205:143

Lung disease, Bronchiolitis oblitherans, 204:133

Lung pathology, diacetyl effects, 204:134

M

Macro-molecular binding, formaldehyde, 203:111

Macromolecular damage, chromium, 210:12

Macrophage function, aging effects, 207:117

Magnitude factors, AF context, 209:64

Mammalian fate and toxicity, fenamiphos, 205:142

Mammalian toxicity, isofenphos, 205:147

Mammals of San Francisco Bay, methyl mercury contamination, 206:124

Mammals, known renal toxicants (table), 207:99–100

Manufacturing emissions, pharmaceuticals, 202:72

Manufacturing sources, perfluoroalkyl isomers, 208:115

Marine environment, biodegradation of oil spills, 206:103

Marine environment, dispersion of oil spills, 206:102

Marine environment, fate of oil spills (diag.), 206:101

Marine environment, oil pollution, 206:98

Marine environment, oil spill effects, 206:97

Marine environment, sediment-oil interaction, 206:103

Marine oil spills, fate in environment, 206:100

Marine organisms, petroleum hydrocarbon accumulation, 206:106

Marine species, bioavailability of hydrocarbons, 206:106

Marine species, PFOS toxicity, 202:15

Mechanism of action, diacetyl, 204:136

Mechanism of toxic action, heavy metals, 203:142

Medical exposure, formaldehyde, 203:107

Medical importance, *Pseudomonas* spp., 201:73

Medication disposal, source of illicit drugs, 210:94

Medicinal plant content, heavy metals (table), 203:141

Medicinal plant potency effects, heavy metals, 203:139 ff.

Medicinal plants, metal effects on metabolism, 203:140

Medicinal plants, metal effects on potency, 203:139 ff.

Medicinal plants, secondary metabolite effects, 203:141

Membrane transport, chemicals, 203:5

Meningitis, *Pseudomonas aeruginosa*, 201:83

Menopausal changes in women, aging effects (table), 207:125

Mercury contamination in fish, San Francisco Bay (diag.), 206:122

Mercury contamination, south San Francisco Bay, 206:121

Mercury residues, south San Francisco Bay sediment cores (diag.), 206:132

Mercury residues, Svalbard glaucous galls, 205:86

Mesocosm data, evaluation & use, 209:102

Mesocosm studies, pesticides, 204:96

Mesocosms, managing mining wastes, 206:34

Metabolic behavior, formaldehyde, 203:106

Metabolic disruption, metals, 210:41

Metabolic effects, aquatic species pesticide uptake, 204:51

Metabolic enzymes, vertebrate effects, 205:93

Metabolic pathways, pesticides (table), 204:53

Metabolism effects, bioaccumulation, 204:48

Metabolism in aquatic species, organophosphate insecticides (table), 204:64

Metabolism in aquatic species, PAHs (table), 204:80

Metabolism in plants, fenamiphos, 205:138

Metabolism of chemicals, aquatic species (table), 204:80

Metabolism of organochlorine pesticides, aquatic species (table), 204:86

Metabolism of pesticides, aquatic species (table), 204:67

Metabolism of pesticides, aquatic species, 204:51, 89

Metabolism of pesticides, enzymology (illus.), 204:57

Metabolism of pharmaceuticals, reactions (table), 202:69

Metabolism, diacetyl, 204:136

Metabolism, OP pesticides, 205:121

Metabolism, pesticides & other chemicals, 204:79

Metabolism, pesticides, 204:1 ff.

Metabolism, pharmaceuticals, 202:68

Metabolism, polycyclic aromatic hydrocarbons, 204:79

Metabolite production, heavy-metal stressed plants (table), 203:143

Metabolites, Svalbard glaucous galls, 205:82

Metal contaminants, phytovolatilization clean-up, 206:38

Metal contaminants, subject to phytostabilization (table), 206:41

Metal contamination, classroom dust (table), 201:5

Metal contamination, dust, 201:6

Metal content, B juncea plant uptake (table), 206:36

Metal effects on secondary metabolites, medicinal plants, 203:141

Metal effects, potency of medicinal plants, 203:139 ff.

Metal excluders, plant responses, 210:13

Metal exposure of humans & wildlife, Red Dog Mine, 206:51

Metal indicators, plant responses, 210:13

Metal toxicants, animal sense disruption, 210:38

Metal toxicity, fish stress, 206:4

Metal toxicity, from iron-ore mining, 206:31

Metal toxicity, gammarid sensitivity, 205:8

Metal toxicity, gammarids, 205:7

Metal toxicity, oxidative stress induced, 203:123

Metal uptake, B. juncea (table), 206:36

Metallothionein induction in gammarids, effects, 205:42

Metallothioneins and metal exposure, biomarkers, 205:49

Metals and metal compounds, pulmonary toxins, 201:54

Metals, behavioral responses, 210:42

Metals, house-dust monitoring, 201:5

Metals, oxidative & metabolic disruption, 210:41

Method comparisons, analyzing pesticide data sets, 209:55

Method for rating quality, single-species aquatic data (tables), 209:27–28

Method for rating quality, aquatic outdoor field data (tables), 209:30–31

Method for rating quality, model ecosystem data (table), 209:30

Method for rating quality, terrestrial lab data (table), 209:31

Methodologies, current AFs used (table), 209:61

Methods comparison, pesticide data analysis (table), 209:55

Methods, for determining octanol-water partitioning coefficients (table), 209:25

Methods, for determining physical-chemical parameters (table), 209:24

Methods, water quality criteria, 209:1 ff.

Methotrexate, skin toxicity, 203:130

Methyl bromide, toxicity, 203:128

Methyl mercury, mammal contamination, 206:124

Methyl salicylate, skin toxicity, 203:130

Methylmercury contamination, San Francisco Bay, 206:123

Methylmercury residue patterns, San Francisco Bay, 206:136

Methylmercury residue patterns, sediments (diag.), 206:135

Methylmercury, bird contamination, 206:123

Microalgae chronic toxicity, PFOS, 202:16

Microbe-dependent bioavailability, naphthalene, 203:29

Microbes uptake, hydrophobic chemicals, 203:24

Microbes, biosurfactant exudate, 203:25

Microbial activity, toxicity assays, 203:29

Microbial bioassays, bioavailability (table), 203:31

Microbial bioavailability, degradation models, 203:47

Microbial bioavailability, OP pesticides, 205:129

Microbial degradation, OP pesticides, 205:127

Microbial degradation, role of ring-hydroxylating oxygenases, 206:66

Microbial effects, chromium, 210:7

Microbial effects, fenamiphos, 205:137

Microbial reductive dechlorination, PCBs (table), 201:145

Microbial remediation, chromium, 210:14

Microbial uptake, chemicals, 203:9

Microcosm data, evaluation & use, 209:102

Microcosm studies, pesticides, 204:96

Microorganisms, chemical degradation, 203:29

Microsomal activity, aging effects, 207:101

Mineral oil effects, marine sea birds, 206:105

Mining slimes, composition (table), 206:33

Mining slimes, iron content (table), 206:33

Mining wastes, characteristics, 206:30

Mining wastes, environmental impact, 206:33

Mining wastes, minimization, 206:33

Mining wastes, remediation approaches, 206:35

Mining wastes, role for mesocom studies, 206:34

Mining wastes, surface runoff, 206:30

Mixed halides, formation mechanism, 208:5

Mixed halides, perfluoroacyl fluoride formation (diag.), 208:6

Mode of action, JHAs, 202:154

Model ecosystem data, evaluation & use, 209:102

Model ecosystem data, quality rating scheme (table), 209:30

Model ecosystem studies, pesticides, 204:92

Model ecosystems, pesticide studies (table), 204:93

Modeling bioavailability, soil fauna, 203:57

Modeling, gammarid feeding activity and rate, 205:19

Models for plants, chemical uptake, 203:51

Models, bioavailability estimation, 203:47, 51

Models, SSD methods, 209:58

Mode-of-action studies, gammarids, 205:32

Mode-of-action test methods, gammarids (table), 205:33

Mode-of-action, OP pesticides, 205:119

Mold, in house dust, 201:18

Molecular chaperones, HSP families, 206:6

Molecular electrostatic potential, diacetyl, 204:139

Money supply, illicit drug contamination, 210:90

Monitoring community-wide health, sewage information mining, 210:98

Monitoring methods, house dust, 201:4

Monitoring pollutants, dust, 201:3

Monod equation, bioavailability connection, 203:47

Morphine sources, environmental contamination, 210:74

Mousse formation, oil spills, 206:101

Multimetric Gammarus spp. tests, perspectives, 205:64

Multi-pathway exposures, aquatic criteria role, 209:14

Multiple stressor biomarkers, Gammarus spp. (table), 205:38

Multispecies data, role in UCDM, 209:100

Multispecies freshwater biomonitor (MFB), biomonitoring system, 205:27

Multispecies use, in criteria derivation, 209:13

Musk residues, fish residues, 207:48

Musk residues, Great Lakes contamination, 207:47

Musk residues, Great Lakes waters, 207:49

Musk residues, water & biota (table), 207:50

Mussel residues, alkyphenol ethoxylates (table), 207:43

Mustard gas, dermatotoxicity, 203:126

Mutagenesis, diacetyl, 204:145

Mutagenicity bioassay, bioavailability, 203:33

Mutagenicity, isofenphos, 205:148

Mutagens, house dust, 201:3

Mycorrhizal fungi, effect on bioavailability, 203:12

N

NAFSAs (N-alkyl-perfluoroalkanesulfon-amides), atmospheric levels, 208:82

NAFSAs, atmospheric concentrations (table), 208:83

NAFSEs (N-alkyl-perfluoroalkanesulfamido-ethanols), atmospheric degradates, 208:60

NAFSEs, atmospheric concentrations (table), 208:87

N-Alkyl-perfluoroalkanesulfamidoethanols
    (NAFSEs), atmospheric lifetime,
    208:59
Nam classification scheme, RHOs (table),
    206:70
Nam classification scheme, RHOs, 206:68
Naphthalene dioxygenases (NDO), shunt
    reaction cycles (diag.), 206:80
Naphthalene, bioluminescence assay, 203:33
Naphthalene, microbe-dependent bioavailabil-
    ity, 203:29
Naphthalene, pulmonary toxin, 201:57
Naphthalenes, toxicity, 203:127
Nearshore Framework Priority, Great Lakes
    monitoring, 207:2
Nemacur, fenamiphos, 205:133
Nematicide description, fenamiphos, 205:133
Nematode infestation vs. PCB levels, Svalbard
    glaucous galls (illus.), 205:100
Nematodes, chemical bioavailability (table),
    203:38
Nerve effects, aging, 207:111
Nervous system, aging effects, 207:109
NEtFOSA (N-ethyl perfluorooctanesulfon-
    amide), isomer separation (illus.),
    208:124
NEtFOSE (N-ethyl
    perfluorooctanesulfonamid-oethanol),
    isomer separation (illus.), 208:124
N-EtFOSE fate, aerobic-activated sludge
    (diag.), 208:169
N-ethyl perfluorooctanesulfonamide
    (NEtFOSA) isomers, separation (illus.),
    208:124
N-ethyl perfluorooctanesulfonamidoethanol
    (NEtFOSE) isomers, separation (illus.),
    208:124
Neurological effects, ACh inhibition, 210:39
Neurological effects, radionuclides, 210:40
Neuronal morphology effects, formaldehyde,
    203:110
Neuro-oxidative effects, formaldehyde,
    203:109
Neurotoxic biomarker, acetylcholinesterase,
    205:50
Neurotoxic effects, OP pesticides, 205:132
Neurotoxic mechanism, OP pesticides,
    205:119
Neurotoxicity, formaldehyde, 203:105 ff.,
    203:108
Neurotransmitters, metal effects, 210:40
Neutrophil function, aging effects, 207:117
n-Hexane, pulmonary toxin, 201:51

Nitric oxide, toxicity, 203:127
Nitroaromatic compounds, pulmonary toxins,
    201:58
Non-Hodgkins lymphoma, toxicants in carpet,
    201:8
Nonylphenol & its ethoxylate, water &
    sediment residues (table), 207:44
Nosocomial infection, Pseudomonas
    aeruginosa, 201:74, 76
Nosocomial pneumonia, bacterial causes
    (table), 201:75
NTPs chemistry generation, pollution
    abatement, 201:119
NTPs chemistry mechanism, pollution
    abatement, 201:119
NTPs chemistry, anti-pollutant applications,
    201:121
NTPs chemistry, bond energies (table),
    201:120
NTPs chemistry, disposal of CBW agents,
    201:128
NTPs chemistry, indoor air purification,
    201:124
NTPs chemistry, origin & definition, 201:118
NTPs chemistry, removing gaseous pollutants,
    201:121
NTPs chemistry, removing odorous pollutants,
    201:122
NTPs chemistry, removing VOC pollutants,
    201:123
NTPs chemistry, solid waste disposal, 201:129
NTPs chemistry, use in sterilization, 201:127
NTPs chemistry, wastewater treatment,
    201:125
NTPs, pollution abatement, 201:117 ff.
Nuclear effects, diatoms, 203:90
Nutrient absorption effects, herbicides, 203:92

O

Occupational exposure studies, Red Dog Mine,
    206:58
Occupational exposure, formaldehyde,
    203:106
Occupational poisoning, OP pesticides,
    205:132
Occupational sources, dermal toxins, 203:130
Occurrence and fate, human pharmaceuticals,
    202:53 ff.
Octanol-water partioning, bioavailability
    estimates, 203:45
Octanol-water partition coefficient, acceptable
    development methods (table), 209:25
Odd fluorotelomer alcohols (oFTOHs),
    atmospheric lifetime, 208:47

oFTOHs (odd fluorotelomer alcohols), atmospheric degradates, 208:47

Oil discharges, Baltic Sea (illus.), 206:109

Oil pollution, marine environment, 206:98

Oil spill effects, dependent factors, 206:104

Oil spill effects, marine environment, 206:97

Oil spill effects, sea-birds, -fish & -animals, 206:104

Oil spill fate, marine environment (diag.), 206:101

Oil spills, Baltic Sea (table), 206:107

Oil spills, causes (diag.), 206:97

Oil spills, emulsification in water, 206:101

Oil spills, evaporation loss, 206:101

Oil spills, fate in marine environment, 206:100

Oil spills, from tanker accidents (table), 206:98

Oil spills, implications, 206:109

Oil spills, international incidence, 206:96

Oil spills, major incidents, 206:97

Oil spills, mousse formation, 206:101

Oil spills, shipping accident implications, 206:95 ff.

Oil-spill biodegradation, marine environment, 206:103

Oil-spill dispersion, marine environment, 206:102

Oil-spill weathering, photo-oxidation, 206:102

OP (organophosphorous) pesticides, description, 205:118

OP mode-of-action, acetylcholinesterase inhibition, 205:119

OP pesticides, acute poisoning, 205:132

OP pesticides, affect of hydrolysis, 205:126

OP pesticides, biological degradation, 205:127

OP pesticides, chemistry, 205:118

OP pesticides, environmental fate and transport, 205:121

OP pesticides, factors affecting bioavailability, 205:129

OP pesticides, health effects, 205:132

OP pesticides, human and wildlife hazards, 205:118

OP pesticides, hydrolysis half lives (table), 205:124

OP pesticides, leaching potential (table), 205:124

OP pesticides, leaching potential, 205:125

OP pesticides, metabolism, 205:121

OP pesticides, microbial bioavailability, 205:129

OP pesticides, microbial degradation, 205:127

OP pesticides, mode-of-action, 205:119

OP pesticides, neurotoxic biomarker, 205:50

OP pesticides, neurotoxic effects, 205:132

OP pesticides, occupational poisoning, 205:132

OP pesticides, oxidation/reduction, 205:126

OP pesticides, photolytic stability, 205:126

OP pesticides, run-off potential, 205:125

OP pesticides, soil adsorption values (table), 205:124

OP pesticides, soil bioavailability, 205:129

OP pesticides, soil half-lives (table), 205:130

OP pesticides, sorption to soil, 205:122

OP pesticides, structural types (illus.), 205:120

OP pesticides, structural types, 205:119

OP pesticides, volatility, 205:124

Organic chemical bioavailability, surfactant effects (table), 203:22

Organic matter, PCB adsorption, 201:147

Organic pollutants in soil, extraction methods, 205:131

Organic wastewater contaminants, hormones & steroids, 207:30

Organic wastewater residues, sediments (table), 207:32–33

Organic wastewater residues, surface waters (table), 207:31

Organic wastewater, chemical pollutants, 207:30

Organisms, chemical uptake, 203:7

Organochlorine contaminants, house dust (table), 201:9, 15

Organochlorine contamination, Svalbard glaucous galls, 205:84

Organochlorine insecticide uptake, soil fauna, 203:35

Organochlorine pesticide bioconcentration, aquatic species (table), 204:24

Organochlorine pesticide metabolism, aquatic organisms (table), 204:86

Organochlorine pesticide metabolism, aquatic species, 204:85

Organochlorine pesticides, chemical structures (illus.), 204:100

Organochlorines in glaucous gulls, legacy chiral forms, 205:81

Organochlorines in house dust, leukemia, 201:11

Organochlorines, gull residues over time (diag.), 205:85

Organochlorines, legacy contaminants, 205:79

Organochlorines, site-specific accumulation effects, 205:88

Organohalogen contaminants, male glaucous gulls (illus.), 205:80

Organometals, Svalbard glaucous galls, **205**:83

Organophosphate degradation, effect of phosphatases, **203**:4

Organophosphate insecticide metabolism, aquatic species (table), **204**:64

Organophosphorous (OP) pesticides, fate and toxicity, **205**:117 ff.

Organophosphorus pesticide bioconcentration, aquatic species (table), **204**:30

Organophosphorus pesticides, aquatic species metabolism, **204**:85

Organophosphorus pesticides, chemical structures (illus.), **204**:101

Organotins, Svalbard glaucous galls, **205**:84

Osteomyelitis, *Pseudomonas aeruginosa*, **201**:77

Osteoporosis induction, alcohol and tobacco, **207**:109

Oxidases, pesticide metabolism, **204**:54

Oxidases, substrates, inducers & inhibitors (illus.), **204**:57

Oxidation/reduction, OP pesticides, **205**:126

Oxidative disruption, metals & radionuclides, **210**:41

Oxidative processes, possible unifying toxic mechanism, **204**:136

Oxidative processes, pulmonary toxicity mechanism, **204**:136

Oxidative stress effects, phenol & quinone, **203**:124

Oxidative stress in plants, induced by chromium, **210**:11

Oxidative stress, antioxidant enzyme relief, **206**:5

Oxidative stress, antioxidant mitigation, **206**:4

Oxidative stress, aquatic pollution, **206**:4

Oxidative stress, asthma & illness, **201**:59

Oxidative stress, cause for COPD, **201**:59

Oxidative stress, dermal toxicity, **203**:119 ff.

Oxidative stress, electron transfer, **204**:134

Oxidative stress, metal exposure, **205**:49

Oxidative stress, pulmonary toxicity, **201**:42

Oxidative stress, toxic action, **204**:135

Oxychlordane effects, gull breeding affects (illus.), **205**:104

Oxygenases, aromatic pollutant degradation, **206**:65 ff.

Oxygenases, Rieske-type proteins, **206**:67

Oxygen-mediated effects, dermal toxicity, **203**:120

Ozone, pulmonary toxin, **201**:45

P

P450 inducers, HSP implications, **206**:7

PAH (polyaromatic hydrocarbon) concentrations, house dust (table), **201**:12, 15

PAH (polycyclic aromatic hydrocarbon) metabolism, aquatic organisms (table), **204**:80

PAHs (polyaromatic hydrocarbons), gammarid toxicity, **205**:8

PAHs (polycyclic aromatic hydrocarbons), dermal toxicity, **203**:125

PAHs , metabolism, **204**:79

PAHs and house dust, cancer, **201**:11

PAHs in house dust, coal burning, **201**:13

PAHs, carpet contamination, **201**:8

PAHs, chemical structures (illus.), **204**:99

PAHs, house dust, **201**:3

PAHs, pulmonary toxins, **201**:57

PAHs, skin cancer, **203**:125

Paint thinner, pulmonary toxin, **201**:51

Paraquat, pulmonary toxin, **201**:56

Parasite effects, gammarid feeding ecology, **205**:21

Parasites, Svalbard glaucous galls, **205**:98

Parathyroid hormone, aging effects, **207**:123

Particulates, pulmonary toxins, **201**:54

Passive diffusion, chemical transport, **203**:5

Passive diffusion, mechanism, **203**:14

Pathogens, pseudomonads, **201**:72

PBBs (polybrominated biphenyls), Svalbard glaucous galls, **205**:81

PBDE (poly brominated diphenyl ethers) contamination, San Francisco Bay, **206**:126

PBDE (polybrominated diphenyl ether) residues, Svalbard glaucous galls, **205**:86

PBDE fish residues, Great Lakes, **207**:61

PBDE residues, Great Lakes biota & sediment (table), **207**:67

PBDE sampling locations, Great Lakes biota (illus.), **207**:60

PBDE sampling locations, Great Lakes sediment (illus.), **207**:60

PBDE, avian species residues, **207**:63

PBDEs, Svalbard glaucous galls, **205**:81

PBDEs, vertebrate metabolism, **205**:82

PCB (polychlorinated benzenes) contamination, San Francisco Bay, **206**:125

PCB (polychlorinated biphenyl) levels vs. nematode infestation, Svalbard glaucous galls (illus.), **205**:100

PCB (polychlorinated biphenyl), volatilization, environmental entry, **201**:140

PCB (polychlorinated byphenyls) contaminants, European Arctic, **205**:79

PCBs carpet contamination, **201**:8

PCBs, adsorption to organic matter, **201**:147

PCBs, aerobic degradation pathway (diag.), **201**:143

PCBs, anaerobic degradation pathway (diag.), **201**:143

PCBs, bioaccumulation, bioconcentration, biomagnification, **201**:147

PCBs, bioaccumulation-affecting factors (table), **201**:151

PCBs, biodegradation and transformation, **201**:142

PCBs, determining environmental fate, **201**:141

PCBs, distribution & environmental fate, **201**:137 ff.

PCBs, effects of aged residues, **201**:139

PCBs, environmental fate processes (diag.), **201**:142

PCBs, factors affecting degradation (diag.), **201**:142

PCBs, global distribution, **201**:139

PCBs, microbial reductive dechlorination (table), **201**:145

PCBs, occurrence and production, **201**:138

PCBs, physicochemical properties (diag.), **201**:139

PCBs, sources of entry (diag.), **201**:140

PCBs, thyroxin ratio effects (illus.), **205**:95

PCBs, uses, **201**:138

PCBs, vertebrate metabolism, **205**:82

PCBs, volatilization, **201**:146

Penicillins, antibiotics, **202**:64

Percentile factors, pesticide data sets (table), **209**:65

Perfluorinated acid precursors, atmospheric occurrence, **208**:1 ff.

Perfluorinated acid precursors, transformation pathways (diag.), **208**:13

Perfluorinated acids (PFAs), from volatile precursors (table), **208**:93

Perfluorinated alcohols, stability, **208**:7

Perfluorinated aldehyde (PFAL) hydrate mechanism, PFCA formation (diag.), **208**:11

Perfluorinated carboxylates (PFA), carbon number effect (diag.), **202**:40

Perfluorinated chain length, PFCA yield, **208**:7

Perfluorinated chemicals, aquatic toxicology, **202**:1 ff.

Perfluorinated chemicals, drinking water residues (table), **208**:200

Perfluorinated compound contamination, Great Lakes watershed, **207**:55

Perfluorinated compounds (PFCs), acronyms (table), **208**:2

Perfluorinated compounds (PFCs), description, **202**:1

Perfluorinated compounds, risk assessments, **207**:55

Perfluorinated compounds, biota residues, **207**:54

Perfluorinated compounds, in air, **208**:204

Perfluorinated compounds, in house dust, **208**:204

Perfluorinated compounds, trophic magnification, **207**:53

Perfluorinated compounds, wildlife contamination, **207**:55

Perfluorinated fatty acids (PFFAs), description, **202**:2

Perfluorinated radical mechanism, atmospheric oxidation, **208**:5

Perfluorinated radicals, PFCA formation (diag.), **208**:6

Perfluorinated substances, fish residues (table), **208**:187

Perfluorinated substances, human exposure, **208**:179 ff.

Perfluorinated substances, in human food, **208**:179 ff.

Perfluorinated sulfonates (PFAS), carbon number effect (diag.), **202**:40

Perfluorinated surfactant contamination, Great Lakes, **207**:52

Perfluorinated surfactant residues, Great Lakes biota (table), **207**:58

Perfluorinated surfactant residues, Great Lakes water (table), **207**:57

Perfluorinated surfactant sampling locations, Great Lakes (illus.), **207**:56

Perfluorinated surfactants, environmental contamination, **207**:51

Perfluorinated surfactants, persistence, **208**:162

Perfluoroacyl fluoride formation, mixed halides (diag.), **208**:6

Perfluoroacyl fluorides, fate, **208**:4

Perfluoroacyl fluorides, hydrolysis lifetime,
    208:4
Perfluoroacyl fluorides, photolytic lifetime,
    208:5
Perfluoroacyl halide formation, hydrofluo-
    roolefin mechanism, 208:7
Perfluoroacyl halides, atmospheric oxidation
    (diag.), 208:8
Perfluoroacyl peroxy radicals, PFCA formation
    (diag.), 208:9
Perfluoroacyl radicals, fate (table), 208:44
Perfluoroalkanesulfonamides, atmospheric
    fate, 208:56
Perfluoroalkyl carboxylates, formulas (table),
    208:113
Perfluoroalkyl isomers, analytical methods,
    208:123
Perfluoroalkyl isomers, manufacturing sources,
    208:115
Perfluoroalkyl isomers, physical-chemical
    effects, 208:131
Perfluoroalkyl sulfonamides, formulae (table),
    208:113
Perfluoroalkyl sulfonates, chromatograms
    (illus.), 208:117
Perfluoroalkyl sulfonates, formulas (table),
    208:113
Perfluorobutanesulfonate (PFBS), toxicity,
    202:22
Perfluorocarboxylate isomers, environmental
    residues, 208:148
Perfluorocarboxylic acids (PFCAs), PFA class,
    208:2
Perfluorooctanesulfonate (PFOS), structure
    (diag.), 202:2
Perfluorooctanesulfonic acid (PFOS), isomer
    composition (table), 208:116
Perfluorooctanoate (PFOA), structure (diag.),
    202:2
Perfluorooctanoic acid (PFOA), isomer
    composition (table), 208:116
Perfluorooctylsulfonamides, structure (diag.),
    202:2
Perfluorosulfonamides, atmospheric sources,
    208:77
Perfluorosulfonamides, chlorine atom kinetics
    (table), 208:57
Perfluorosulfonamides, hydroxyl radical
    kinetics (table), 208:57
Perfluorosulfonic acid (PFSA) formation, from
    perfluorosulfonamides, 208:11
Perfluorosulfonic acids, PFA class, 208:2

Persistence in soil, OP pesticides (table),
    205:130
Persistence, perfluorinated surfactants, 208:162
Personal care products, Great Lakes waters,
    207:35
Persulfate & perchlorate, pulmonary toxins,
    201:55
Pesticide bioaccumulation, aquatic organisms,
    204:1 ff.
Pesticide bioaccumulation, aquatic species,
    (table), 204:43
Pesticide bioaccumulation, theory, 204:49
Pesticide bioconcentration, age dependence,
    204:14
Pesticide bioconcentration, aquatic organisms,
    204:1 ff.
Pesticide bioconcentration, aquatic species
    (table), 204:35
Pesticide bioconcentration, environmental
    effects, 204:16
Pesticide biota-sediment accumulation factors,
    aquatic species (table), 204:46
Pesticide carpet contamination, track in, 201:7
Pesticide class, BCF effect, 204:42
Pesticide clearance times, BCF effects, 204:23
Pesticide contaminants, San Francisco Bay,
    206:128
Pesticide data analysis, methods comparison
    (table), 209:55
Pesticide data set analysis, method
    comparisons, 209:55
Pesticide data sets, acute toxicity (table),
    209:39
Pesticide data sets, Burr Type III distribution
    fit (table), 209:51
Pesticide data sets, log-normal distribution fit
    (table), 209:51
Pesticide data sets, log-triangular distribution
    fit (table), 209:51
Pesticide data sets, percentile factors (table),
    209:65
Pesticide detection, Great Lakes, 207:14
Pesticide detections, Great Lakes water, 207:15
Pesticide effects, freshwater diatoms,
    203:87 ff.
Pesticide effects, gammarid reproduction,
    205:44
Pesticide elimination, species variability,
    204:40
Pesticide exposure types, gammarids, 205:53
Pesticide exposure, banned pest products,
    201:8
Pesticide lawn use, home contamination, 201:7

Pesticide metabolic pathways, aquatic species
(table), **204**:53

Pesticide metabolism in aquatic species,
esterases, **204**:63

Pesticide metabolism, aquatic organisms,
204:1 ff.

Pesticide metabolism, aquatic species (table),
204:67

Pesticide metabolism, aquatic species
differences, **204**:90

Pesticide metabolism, aquatic species, **204**:51

Pesticide metabolism, aquatic species, **204**:89

Pesticide metabolism, bioconcentration effects,
204:14

Pesticide metabolism, enzyme role, **204**:52

Pesticide metabolism, enzymology (illus.),
204:57

Pesticide metabolism, oxidases, **204**:54

Pesticide metabolism, sulfotransferases,
204:73

Pesticide monitoring, lawn care & agricultural
chemicals, **207**:14

Pesticide residue levels, Great Lakes (illus.),
207:18

Pesticide sampling locations, Great Lakes
(illus.), **207**:13

Pesticide sensitivity of diatoms, interfering
factors, **203**:96

Pesticide studies, micro- & mesocosms, **204**:96

Pesticide studies, model ecosystems (table),
204:93

Pesticide toxicity, gammarids, **205**:7

Pesticide toxicity, multiple-component effects,
203:88

Pesticide uptake, aquatic organism growth
effects, **204**:14

Pesticide volatility, ingestion vs. inhalation,
201:7

Pesticides & chemicals, metabolism, **204**:79

Pesticides & other chemicals, structures
(illus.), **204**:102

Pesticides in aquatic species, BCF vs. log K_{ow}
(diag.), **204**:41

Pesticides in fish, BCF vs. log K_{ow} (diag.),
204:7

Pesticides listed in this volume, chemical
names (table), **209**:128

Pesticides, aquatic environmental quality,
203:88

Pesticides, Burr III family distribution fit
(table), **209**:50

Pesticides, chemical structures (illus.), **204**:103

Pesticides, factors affecting bioconcentration,
204:3

Pesticides, farm-worker exposure, **201**:7

Pesticides, fate and toxicity of OPs, **205**:117 ff.

Pesticides, in house dust, **201**:6

Pesticides, model ecosystem studies, **204**:92

Pesticides, physico-chemical effects on BCF
(table), **204**:5

Pesticides, plant sorption, **203**:34

Pesticides, water quality criteria, **209**:1 ff.

Petroleum constituents, crude oil, **206**:98

Petroleum hydrocarbon accumulation, marine
organisms, **206**:106

Petroleum products, sea water pollution,
206:95

Petroleum products, total consumption (table),
206:96

Petroleum transport, ship tankers, **206**:96

PFA (perfluorinated acid) classes, PFSAs
& PFCAs, **208**:2

PFA formation, atmospheric mechanism, **208**:4

PFA isomers, bioaccumulation, **208**:149

PFA isomers, branching effects, **208**:133

PFA isomers, nomenclature, **208**:112

PFA isomers, separation methods, **208**:125

PFA isomers, toxicity differences, **208**:149

PFA levels, precursors impact, **208**:82

PFA precursors, anthropogenic products, **208**:2

PFA precursors, atmospheric occurrence,
208:1 ff.

PFA precursors, chemistry, **208**:12

PFA precursors, chlorofluorocarbon
replacements, **208**:2

PFA precursors, nomenclature, **208**:112

PFA sources, environmental loading effects,
208:122

PFA structure effects, environmental behavior,
208:112

PFA substances, isomer profiling, **208**:111 ff.

PFAL (perfluorinated aldehyde) hydrate
mechanism, PFCA formation (diag.),
208:11

PFAL hydrates, atmospheric lifetime, **208**:35

PFALs, atmospheric degradates, **208**:36, 42

PFALs, photolysis properties (table), **208**:43

PFALs, photolysis, **208**:42

PFAs (perfluorinated acids), from volatile
precursors (table), **208**:93

PFAS (perfluorinated sulfonates), carbon
number effect (diag.), **202**:40

PFAS, aquatic species toxicity (diag.), **202**:43

PFAs, environmental impact, **208**:1 ff.

PFAs, major classes, **208**:2

PFAs, sources, transport & yield, **208**:94

PFAs, volatile precursors (table), **208**:93

PFASs (per- and poly-fluorinated alkyl substances), Svalbard glaucous gall residues, **205**:83

PFBS (perfluorobutanesulfonate), toxicity, **202**:22

PFBS, aquatic toxicity (table), **202**:23

PFBS, avian TRVs (toxicity reference values) (table), **202**:37

PFBS, TRVs, **202**:37

PFC (perfluorinated compound) contamination, patterns & associations, **208**:207

PFC levels, aquatic invertebrates (table), **208**:189

PFC precursors, formulae (table), **202**:3

PFC residues, fish & seafood, **208**:180

PFCA (perfluorinated carboxylates), carbon number effect (diag.), **202**:40

PFCA (perfluorocarboxylic acid) formation, atmospheric mechanism, **208**:4

PFCA formation, from perfluorinated radicals (diag.), **208**:6

PFCA formation, perfluorinated aldehyde hydrate mechanism (diag.), **208**:11

PFCA formation, via perfluoroacyl peroxy radicals (diag.), **208**:9

PFCA yield, perfluorinated chain length, **208**:7

PFCA, aquatic species toxicity (diag.), **202**:43

PFCAs sources & characteristics, **208**:97

PFCAs, atmospheric formation, **208**:2

PFCAs, direct atmospheric formation, **208**:8

PFCs (perfluorinated compounds), description, **202**:1

PFCs, aquatic species water quality criteria values (diag.), **202**:25

PFCs, direct food contamination, **208**:196

PFCs, drinking water contamination, **208**:199

PFCs, drinking water residues (table), **208**:200

PFCs, edible fish residues (table), **208**:181

PFCs, environmental contaminants, **208**:179

PFCs, food coating uses, **208**:195

PFCs, food contamination, **208**:195

PFCs, formulae (table), **202**:3

PFCs, house dust contamination (table), **208**:205

PFCs, in edible invertebrates, **208**:193

PFCs, indirect food contamination, **208**:195

PFCs, QSAR analysis, **202**:39

PFCs, safety limits & TDIs, **208**:202

PFCs, trophic level effects, **208**:193

PFCs, water quality criteria, **202**:22

PFFAs (perfluorinated fatty acids), description, **202**:2

PFFAs, ecotoxicity, **202**:11

PFFAs, target organs, **202**:4

PFOA (perfluorooctanoate), structure (diag.), **202**:2

PFOA (perfluorooctanoic acid), isomer composition (table), **208**:116

PFOA chromatograms, isomer profiles (illus.), **208**:120

PFOA isomer content, environmental pattern, **208**:121

PFOA isomer profiles, environmental residues, **208**:137

PFOA isomers in animals, structure-property effects (illus.), **208**:152

PFOA isomers, environmental residues (table), **208**:134

PFOA isomers, in biological samples (table), **208**:134

PFOA, fish residues, **208**:180

PFOA, GMAV and GMCV (table), **202**:32

PFOA, house dust contamination (table), **208**:205

PFOA, in Japanese houses, **208**:204

PFOA, isomer composition (table), **208**:132

PFOAs, analytical biases (illus.), **208**:127

PFOAs, analytical biases, **208**:126

PFOAs, distribution in humans (table), **208**:128

PFOS (perfluorooctane sulfonate), Svalbard glaucous gall residues, **205**:83

PFOS (perfluorooctanesulfonate), structure (diag.), **202**:2

PFOS (perfluorooctanesulfonic acid), isomer composition (table), **208**:116

PFOS bioconcentration, fish, **202**:10

PFOS chromatograms, isomer profiles (illus.), **208**:120

PFOS chronic toxicity, amphibians, **202**:20

PFOS chronic toxicity, aquatic macrophytes, **202**:19

PFOS chronic toxicity, invertebrates, **202**:20

PFOS derivatives, biodegradation, **208**:168

PFOS fluoride-derived products, isomer profiles (table), **208**:121

PFOS in fish, critical body burdens, **202**:28

PFOS isomer composition, environmental media (table), **208**:144

PFOS isomer composition, humans (table), **208**:140

PFOS isomer composition, wildlife (table), **208**:144

PFOS isomer profiles, organisms, **208**:139

PFOS isomers in animals, structure-property effects (illus.), **208**:152

PFOS residues, fish advisory, **208**:194

PFOS toxicity, acute aquatic (table), **202**:12

PFOS toxicity, amphibians, **202**:15

PFOS toxicity, fish, **202**:15

PFOS toxicity, marine species, **202**:15

PFOS, acute aquatic toxicity (table), **202**:12

PFOS, aquatic species chronic toxicity (table), **202**:17

PFOS, BCF in fish (table), **202**:9

PFOS, bioaccumulation, **202**:8

PFOS, bioconcentration, **202**:8

PFOS, biodegradation, **202**:6

PFOS, biomagnification, **208**:194

PFOS, bluegill cumulative mortality (table), **202**:29

PFOS, chemical character, **202**:5

PFOS, chronic toxicity in fish, **202**:21

PFOS, current production, **208**:121

PFOS, degradability, **208**:165

PFOS, dominant PFC fish contaminant, **208**:188

PFOS, enterohepatic recirculation, **202**:8

PFOS, environmental fate, **202**:5

PFOS, FAVs (table), **202**:27

PFOS, fish residues, **208**:180

PFOS, house dust contamination (table), **208**:205

PFOS, hydrolysis, **202**:6

PFOS, in Japanese houses, **208**:204

PFOS, invertebrate toxicity, **202**:14

PFOS, microalgae chronic toxicity, **202**:16

PFOS, photolysis, **202**:6

PFOS, phys-chemical properties (table), **202**:6

PFOS, salt forms, **202**:5

PFOS, thermal stability, **202**:7

PFOS, toxicity reference values (TRV) (table), **202**:34

PFOSs, analytical biases, **208**:126

PFOSs, isomer separation (illus.), **208**:130

PFSA (perfluorosulfonic acid) formation, from perfluorosulfonamides, **208**:11

pH effects, criteria compliance, **209**:121

Phagocytosis (endocytosis), chemical transport, **203**:7

Pharmaceutical contamination, environmental pathways (diag.), **202**:71

Pharmaceutical counterfeiting, illicit drugs, **210**:68

Pharmaceutical detection, sewage treatment plants (table), **202**:74–79

Pharmaceutical detections, Great Lakes, **207**:23

Pharmaceutical detections, rural streams, **207**:23

Pharmaceutical detections, sewage sludge (table), **202**:104–108

Pharmaceutical detections, soil (table), **202**:109

Pharmaceutical detections, surface water (table), **202**:79–84

Pharmaceutical detections, water & sediment, **207**:25

Pharmaceutical fate, wastewater treatment plants, **202**:111

Pharmaceutical release, drug disposal, **202**:73

Pharmaceutical release, sludge land-use, **202**:73

Pharmaceutical release, wastewater irrigation, **202**:73

Pharmaceutical removal, drinking water (table), **202**:112–117

Pharmaceutical residue levels, surface waters (illus.), **207**:28

Pharmaceuticals in water samples, detection frequency, **207**:27

Pharmaceuticals metabolism, reactions (table), **202**:69

Pharmaceuticals sold, volume by country (table), **202**:56

Pharmaceuticals, environmental contaminants, **207**:19

Pharmaceuticals, abiotic degradation, **202**:127

Pharmaceuticals, anemia association (table), **207**:108

Pharmaceuticals, beta-blockers, **202**:66

Pharmaceuticals, biodegradability (table), **202**:120–123

Pharmaceuticals, drinking water detections (table), **202**:90–95

Pharmaceuticals, ecotoxicology studies, **202**:54

Pharmaceuticals, environmental contamination, **202**:74

Pharmaceuticals, environmental fate, **202**:110

Pharmaceuticals, environmental release, **202**:72

Pharmaceuticals, excretion rates (table), **202**:70

Pharmaceuticals, fate in drinking water, **202**:134

Pharmaceuticals, fate in soil, **202**:129, 132

Pharmaceuticals, Great Lakes sampling locations (illus.), **207**:23

Pharmaceuticals, groundwater detections
    (table), **202**:85–89
Pharmaceuticals, hepatotoxic agents (table),
    207:103–104
Pharmaceuticals, listing (table), **202**:137–143
Pharmaceuticals, metabolism, **202**:68
Pharmaceuticals, occurrence and fate,
    202:53 ff.
Pharmaceuticals, photodegradation (tables),
    202:127–128
Pharmaceuticals, production emissions, **202**:72
Pharmaceuticals, sediment detections (table),
    202:96–97
Pharmaceuticals, sediment residues, **207**:28
Pharmaceuticals, sediment sorption, **202**:123
Pharmaceuticals, sewage treatment plants,
    202:72
Pharmaceuticals, sorption (table), **202**:124–126
Pharmaceuticals, structures & properties
    (table), **202**:57–62
Pharmaceuticals, surface water residues
    (table), **207**:26–27
Pharmaceuticals, use and consumption, **202**:55
Pharmacology, chemical-drug interactions,
    207:131
pH-dependency, bioconcentration, **204**:17
Phenol, oxidative stress effects, **203**:124
Phenols, pulmonary toxins, **201**:52
Phosgene, pulmonary toxin, **201**:47
Photodegradation, pharmaceuticals (tables),
    202:127–128
Photolysis properties, flurotelomer aldehydes
    (table), **208**:43
Photolysis properties, PFALs (table), **208**:43
Photolysis, perfluoroacyl fluorides, **208**:4
Photolysis, PFALs, **208**:42
Photolysis, PFOS, **202**:6
Photolytic pathway, fenoxycarb (diag.),
    202:170
Photolytic stability, OP pesticides, **205**:126
Photo-oxidation, oil-spill weathering, **206**:102
Photosynthesis inhibition in algae, herbicides,
    203:91
Photosynthesis inhibition, chromium, **210**:10
Phthalates detections, Great Lakes waters,
    207:36
Phthalates, house dust contaminant, **201**:13
Phthalates, pulmonary toxins, **201**:57
Physical properties, fenamiphos (table),
    205:134
Physical properties, PFOS (table), **202**:6
Physical-chemical data evaluation, UCDM,
    209:101

Physical-chemical data generation, acceptable
    methods (table), **209**:24
Physical-chemical data, differential quality,
    209:22
Physical-chemical data, water quality criteria
    use, **209**:22
Physico-chemical properties, chromium
    (table), **210**:3
Physicochemical properties, fenoxycarb
    (table), **202**:160
Physicochemical properties, PCBs (diag.),
    201:139
Physiological adaptations of stressed fish,
    HSPs, **206**:1 ff.
Physiological mechanism, behavioral link,
    210:37
Phyto- (green)-remediation, chromium, **210**:16
Phytoaccumulation of metals, *B. juncea*
    (table), **206**:36
Phytoaccumulator plants, for phytoremedia-
    tion, **206**:35
Phytodetoxification, phytoremediation process,
    210:22
Phytoextraction processes, soil, plant & energy
    recovery (diag.), **210**:19
Phytoextraction, chelate-assisted uptake,
    210:20
Phytoextraction, natural plant uptake, **210**:20
Phytoextraction, phytoremediation approach,
    206:37
Phytoextraction, phytoremediation method,
    210:18
Phytoextraction, representation (illus.), **210**:18
Phytoremediation approach, phytoextraction,
    206:37
Phytoremediation improvement, genetic
    engineering, **210**:23
Phytoremediation method, phytoextraction,
    210:18
Phytoremediation process, phytodetoxification,
    210:22
Phytoremediation process, phytostabilization,
    206:40
Phytoremediation process, phytostabilization,
    210:22
Phytoremediation process, phytovolatilization,
    206:38
Phytoremediation process, phytovolatilization,
    210:21
Phytoremediation process, rhizofiltration,
    206:39
Phytoremediation process, rhizofiltration,
    210:22

Phytoremediation, characterization, 206:35
Phytoremediation, clean mining wastes, 206:31
Phytoremediation, contaminant removal,
    206:87
Phytoremediation, iron-ore wastes, 206:29 ff.
Phytoremediation, lemon grass (illus.), 206:35
Phytoremediation, suitable plant species,
    206:42
Phytoremediation, sustainable remediation
    approach, 206:34
Phytoremediation, tomato plants (illus.),
    206:36
Phytoremediation, tree species (illus.), 206:36
Phytostabilization, candidate plant species
    (table), 206:41
Phytostabilization, depiction (illus.), 210:23
Phytostabilization, mechanism depicted
    (illus.), 206:40
Phytostabilization, phytoremediation process,
    206:40
Phytostabilization, phytoremediation process,
    210:22
Phytotoxic effects, chromium, 210:12
Phytotoxic symptoms, chromium, 210:9
Phytotoxicity, chromium, 210:9
Phytovolatilization, depiction (illus.), 210:21
Phytovolatilization, phytoremediation process,
    206:38
Phytovolatilization, phytoremediation process,
    210:21
Plant bioassays, bioavailability measurement
    (table), 203:36
Plant bioavailability, uptake models, 203:51
Plant cell signaling, heavy metal effects (diag.),
    203:144
Plant degradation, isofenphos, 205:146
Plant effects, heavy metal exposure, 210:13
Plant effects, heavy metals, 203:139
Plant growth effects, heavy metals, 203:140
Plant growth retardation, chromium, 210:10
Plant metabolism, fenamiphos, 205:138
Plant phytoextraction, phytoremediation
    approach, 206:37
Plant response strategies, heavy metals (illus.),
    210:13
Plant responses, hyperaccumulators, 210:13
Plant responses, metal excluders, 210:13
Plant responses, metal indicators, 210:13
Plant roots, chemical uptake, 203:9
Plant species candidates, phytostabilization
    (table), 206:41
Plant species, capable of phytoremediation,
    206:35

Plant species, hyperaccumulators, 206:42
Plant species, suitable for phytoremediation,
    206:42
Plant uptake, natural phytoextraction, 210:20
Plants, chemical uptake, 203:34
Plants, chromium accumulation, 210:5
Plants, chromium contamination, 210:5
Plants, chromium-induced enzyme changes,
    210:11
Plants, chromium-induced oxidative stress,
    210:11
Plants, hyperaccumulators of mining waste
    (table), 206:39
PLANTX model, chemical uptake in plants,
    203:52
Plasticizer residues, surface waters (table),
    207:31
Pleopod beat frequency, pollutant bioassay,
    205:29
PNDEs (polybrominated diphenyl ethers),
    house dust contamination (table),
    201:15
Pneumonia, Pseudomonas aeruginosa, 201:78
Pollutant abatement, NTPs chemistry
    mechanism, 201:119
Pollutant bioassay, pleopod beat frequency,
    205:29
Pollutant degradation, oxygenases, 206:65 ff.
Pollutant effects in gammarids, embryotoxicity,
    205:42
Pollutant extraction, by carpet vacuuming,
    201:22
Pollutant problems in homes, discussion,
    201:28
Pollutant protective response, HSP production,
    206:10
Pollutant regulation, different countries (table),
    207:7–11
Pollutants abatement, NTPs chemistry,
    201:117 ff.
Pollutants and disease, infants, 201:2
Pollutants in Canadian waters, current-use
    pesticides (table), 207:16–17
Pollutants in house dust, exposure, 201:2
Pollutants in house dust, infants, 201:1 ff.
Pollutants in the Great Lakes, current-use
    pesticides (table), 207:16–17
Pollutants, home cleaning, 201:19
Pollution abatement tool, NTPs chemistry,
    201:118
Pollution biomarkers, antioxidants, 206:5
Pollution effects, algae, 203:88
Pollution sources, formaldehyde, 203:106

Pollution, oil spill causes (diag.), **206**:97

Polyaromatic hydrocarbons (PAHs) detections, Great Lakes waters, **207**:36

Polybrominated diphenyl ethers (PBDE), environmental contamination, **207**:59

Polycyclic aromatic hydrocarbons (PAHs), sediment concentrations from oil (table), **206**:104

Polypharmacy, adverse drug interactions, **207**:132

Polysaccharide inhibition in algae, herbicides, **203**:92

Popcorn lung disease, diacetyl toxicity, **204**:133 ff.

Popcorn lung disease, electron transfer, **204**:133 ff.

Popcorn plant workers, Bronchiolitis oblitherans, **204**:133

Population dynamics effect, gammarids, **205**:43

Population surveys vs. FEUDS, disadvantages, **210**:85

Population-level effects, key endpoints, **209**:12

Population-level endpoints, use in UCDM, **209**:13

Porphyrins, Svalbard glaucous galls, **205**:93

Post-exposure feeding depression, *Daphnia*, **205**:20

Potable water, *Pseudomonas aeruginosa* contamination (table), **201**:88

Predation effects, gammarids, **205**:20

Production, chromium, **210**:5

Prolactin effects, Svalbard glaucous gall residues, **205**:96

Properties and structures, pharmaceuticals (table), **202**:57–62

Protein glycation, diacetyl & other α-dicarbonyls, **204**:142

Protein synthesis inhibition in algae, herbicides, **203**:92

Proteins, metabolism and HSPs, **206**:6

Proteins, role in stressed fish, **206**:7

Pseudomonad classification, genera (table), **201**:73

Pseudomonads, bacteria group, **201**:72

Pseudomonas aeruginosa contamination, water, **201**:87–91

Pseudomonas aeruginosa illness, pools & tap water, **201**:99

Pseudomonas aeruginosa illness, spas & tubs (table), **201**:98

Pseudomonas aeruginosa, AIDS patients, **201**:85

Pseudomonas aeruginosa, aqueous media survival (table), **201**:94

Pseudomonas aeruginosa, biofilms, **201**:92

Pseudomonas aeruginosa, cancer & granulocytopenic patients, **201**:86

Pseudomonas aeruginosa, disinfection, **201**:96

Pseudomonas aeruginosa, drinking water risks (table), **201**:104

Pseudomonas aeruginosa, ear & eye infections, **201**:81

Pseudomonas aeruginosa, healthy-human infections (table), **201**:75

Pseudomonas aeruginosa, human health effects, **201**:75

Pseudomonas aeruginosa, immunocompromised humans (table), **201**:76

Pseudomonas aeruginosa, in drinking water, **201**:87

Pseudomonas aeruginosa, infection of children, **201**:83

Pseudomonas aeruginosa, infective dose, **201**:102

Pseudomonas aeruginosa, lung infection & cystic fibrosis, **201**:84

Pseudomonas aeruginosa, meningitis, **201**:83

Pseudomonas aeruginosa, nosocomial disease, **201**:76

Pseudomonas aeruginosa, nosocomial infection, **201**:74

Pseudomonas aeruginosa, occurrence & survival, **201**:86

Pseudomonas aeruginosa, pneumonia, **201**:78

Pseudomonas aeruginosa, pool contamination (table), **201**:89

Pseudomonas aeruginosa, risk assessment, **201**:101

Pseudomonas aeruginosa, risk in water, **201**:72 ff.

Pseudomonas aeruginosa, septicemia, endocarditis & osteomyelitis, **201**:77

Pseudomonas aeruginosa, skin & burn-wound infections, **201**:80

Pseudomonas aeruginosa, sources, **201**:93

Pseudomonas aeruginosa, survival, **201**:94

Pseudomonas aeruginosa, transmission mode, **201**:84

Pseudomonas aeruginosa, UV disinfection (table), **201**:96

Pseudomonas aeruginosa, water quality standards, **201**:101

Pseudomonas aeruginosa, water transmission, **201**:97

Pseudomonas aeruginosa-induced disease, recreational water (table), **201**:90

Pseudomonas bacteremia, septicemia, **201**:77

Pseudomonas spp., characteristics, **201**:72

Pseudomonas spp., human pathogenicity, **201**:74

Pseudomonas spp., medical importance, **201**:73

Pulmonary injury, ROS, **201**:43

Pulmonary toxicity mechanism, electron transfer, **201**:43

Pulmonary toxicity, antioxidant protection, **201**:41 ff.

Pulmonary toxicity, electron transfer, **201**:41 ff.

Pulmonary toxicity, ROS and oxidative stress, **201**:42

Pulmonary toxicity, unifying mechanism proposal, **204**:136

Pulmonary toxin, ozone, **201**:45

Pulmonary toxins, acrolein, epichlorohydrin, chloroform & carbon tetrachloride, **201**:50

Pulmonary toxins, anesthetics & therapeutic agents, **201**:53

Pulmonary toxins, asbestos, silica, persulfate & perchlorate, **201**:55

Pulmonary toxins, benzene, toluene, styrene, trichloroethylene, **201**:49

Pulmonary toxins, diacetyl, phenols, 1,3-butadiene, 2,4-decadienal, **201**:52

Pulmonary toxins, ethanol, n-hexane, paint thinner, carbon disulfide, **201**:51

Pulmonary toxins, ethylene oxide, sarin & sulfur mustard, **201**:46

Pulmonary toxins, formaldehyde, phosgene, carbon monoxide & hydrogen cyanide, **201**:47

Pulmonary toxins, hydrogen sulfide, sulfur dioxide, chlorine & hypochlorous acid, **201**:48

Pulmonary toxins, metals & particulates, **201**:54

Pulmonary toxins, naphthalene, PAHs, phthalates & quinones, **201**:57

Pulmonary toxins, nitroaromatic compounds & radiation, **201**:58

Pulmonary toxins, pesticides, tobacco, cocaine, **201**:56

Pulmonary toxins, structures (diag.), **201**:44

Pulsed exposure models, gammarid effects, **205**:57

Pyrethroid insecticide bioconcentration, aquatic species (table), **204**:34

Q

QSAR (quantitative structure activity relationships) approach, data-gap filling, **209**:15

QSAR analysis, PFCs, **202**:39

QSAR, interspecies correlations, **209**:16

QSARs, toxicity estimation procedure (table), **209**:126

Quality assurance, FEUDS, **210**:86

Quantitative structure activity relationships (QSAR), filling data gaps, **209**:15

Quinolones, antibiotics, **202**:64

Quinones, oxidative-induced toxicity, **203**:124

Quinones, pulmonary toxins, **201**:57

R

Racing animal drugs, illegal use, **210**:92

Radiation, dermal effects, **203**:123

Radiation, pulmonary toxin, **201**:58

Radicals and pulmonary toxicity, possible mechanism, **201**:41 ff.

Radionuclide effects, animal behavior, **210**:35 ff.

Radionuclide effects, human & ecological consequences, **210**:49

Radionuclides, neurological effects, **210**:40

Radionuclides, oxidative disruption, **210**:41

Ranking scheme, data relevance & reliability (table), **209**:31

Rats, soil toxicity, **203**:41

Reaction cycles, NDO shunts (diag.), **206**:80

Reaction kinetics, HCFC-123, **208**:20

Reactive carbonyl species, toxicity, **203**:128

Reactive nitrogen species, similarity to ROS, **201**:45

Reactive oxygen species (ROS), biological stress, **206**:4

Reactive oxygen species (ROS), cellular injury & ARDS, **201**:43

Reactive oxygen species (ROS), electron transfer, **204**:134

Reactive oxygen species, dermal toxicity, **203**:119 ff.

Reactive oxygen species, dermatotoxicity schematic (diag.), **203**:122

Reactive oxygen species, environmental stress, **206**:1

Reactive oxygen species, skin-damaging effects, **203**:122

Reactive oxygen species, toxic action, **204**:135

Recreational water, *Pseudomonas aeruginosa* contamination & disease, 201:8, 90

Red Dog Mine employees, blood lead levels (illus.), 206:59

Red Dog Mine heavy-metal exposure, Alaska, 206:49

Red Dog Mine, contaminated soil contact, 206:54

Red Dog Mine, dust control, 206:60

Red Dog Mine, dust-borne metal exposure, 206:53

Red Dog Mine, environmental exposure studies, 206:57

Red Dog Mine, epidemiology studies, 206:57

Red Dog Mine, heavy metal exposure, 206:51

Red Dog Mine, heavy metal toxicity, 206:51

Red Dog Mine, heavy-metal human exposure, 206:49 ff.

Red Dog Mine, inhalation of fugitive dust, 206:53

Red Dog Mine, lead exposure effect, 206:56

Red Dog Mine, lead exposure, 206:52

Red Dog Mine, maps (illus.), 206:50

Red Dog Mine, occupational exposure studies, 206:58

Red Dog Mine, regulatory oversight, 206:60

Red Dog Mine, resident mine dust exposure, 206:55

Red Dog Mine, wildlife & human metal exposure, 206:51

Redox cycling and electron transfer, toxicity mechanism, 201:43

Redox cycling, electron transfer, 204:136

Reducing infant exposure, house dust, 201:1 ff.

Regioselectivity, RHOs, 206:81

Regression analysis, ecotoxicity evaluation, 209:9–10

Regulatory standards by chemical category, international pollutants (table), 207:7–11

Regulatory testing, bioconcentration, 204:2

Remediation approaches, mining wastes, 206:35

Renal toxicants, humans & lab animals (table), 207:99–100

Reproduction effects, Svalbard glaucous gall residues, 205:101

Reproductive behavior effects, Svalbard glaucous gall residues, 205:102

Reproductive endpoints, ecotoxicity assessment, 209:11

Reproductive toxicity, chromium, 210:8

Reproductive toxicity, isofenphos, 205:148

Resident exposure, mine dust, 206:55

Residue accumulation affects, Svalbard glaucous galls, 205:88

Residue levels, Great Lakes pesticides (illus.), 207:18

Residue uptake, earthworms, 203:35

Respiration rate, gammarids, 205:32

Respiratory system, aging effects, 207:130

Response addition model, toxicity of mixtures, 209:83

Retinoid effects, Svalbard glaucous gall metabolism, 205:93

Rhizofiltration, phytoremediation process, 206:39

Rhizofiltration, phytoremediation process, 210:22

RHO (ring-hydroxylating oxygenase) active sites, catalytic pockets (illus.), 206:83–84

RHO classification, Batie scheme (table), 206:69

RHO degradation techniques, improvements, 206:85

RHO pockets, structural residues (table), 206:82

RHOs, aromatic pollutant degradation, 206:65 ff.

RHOs, Batie classes, 206:67

RHOs, classification systems, 206:66

RHOs, description, 206:66

RHOs, electron transfer, 206:78

RHOs, Kweon classification scheme (table), 206:71–2

RHOs, Kweon classification scheme, 206:70

RHOs, microbial degradation role, 206:66

RHOs, Nam classification scheme (table), 206:70

RHOs, Nam classification, 206:68

RHOs, regio- & stereo-selectivity, 206:81

RHOs, structural investigations, 206:72

RHOs, substrate oxidation, 206:78

RHOs, typical structures (illus.), 206:74

RHOs, α subunit structure, 206:75–8

RHOs, β subunit structure, 206:73

Rieske proteins, characterization, 206:67

Rieske-type proteins, oxygenases, 206:67

Risk assessment, *Pseudomonas aeruginosa*, 201:101

Risk assessment, *Pseudomonas aeruginosa*, 201:72 ff.

Risk assessments, perfluorinated compounds, 207:55

Rodent behavior studies, uranium, caesium & cadmium, **210**:44

Rodent model, human behavioral response, **210**:44

Root uptake of chemicals, influencing factors, **203**:11

ROS, pulmonary toxicity, **201**:42

Run-off potential, OP pesticides, **205**:125

S

Safer cleaning products, home cleaning, **201**:23

Safety limits, PFCs, **208**:202

Salt pond restoration, San Francisco Bay, **206**:117

Samplers, high volume for pesticides, **201**:4

Sampling and monitoring, house dust, **201**:4

Sampling sites, pesticides in the Great Lakes (illus.), **207**:13

San Francisco Bay mammals, methyl mercury contamination, **206**:124

San Francisco Bay sediment cores, south Bay mercury residues (diag.), **206**:132

San Francisco Bay water quality, biological aspects, **206**:130

San Francisco Bay water quality, chemical aspects, **206**:130

San Francisco Bay, bird contamination, **206**:123

San Francisco Bay, birds & water quality, **206**:137

San Francisco Bay, contaminant sequestration in marshes, **206**:137

San Francisco Bay, dioxin contamination, **206**:127

San Francisco Bay, emerging contaminants, **206**:129

San Francisco Bay, mercury contamination, **206**:121

San Francisco Bay, mercury residues in fish (diag.), **206**:122

San Francisco Bay, methylmercury contamination, **206**:123

San Francisco Bay, methylmercury residue patterns, **206**:136

San Francisco Bay, PBDE (polybrominated diphenyl ether) contamination, **206**:126

San Francisco Bay, PCB (polychlorinated biphenyl) contamination, **206**:124

San Francisco Bay, pesticide contaminants, **206**:128

San Francisco Bay, salt ponds and wetland restoration, **206**:117

San Francisco Bay, south Bay contamination, **206**:120

San Francisco Bay, south Bay Salt Pond Restoration Project (illus.), **206**:119

San Francisco Bay, south Bay sediment erosion, **206**:131

San Francisco Bay, south Bay water quality, **206**:115 ff.

San Francisco Bay, south Bay water quality, **206**:120

San Francisco Bay, tidal marsh habitats (illus.), **206**:118

San Francisco Bay, watershed map (illus.), **206**:117

San Francisco Bay, wildlife residues, **206**:125

San Francisco Estuary, BDE (brominated diphenyl ether) contamination (illus.), **206**:127

San Francisco Estuary, description, **206**:116

Sea animals, oil spill effects, **206**:104

Sea bird effects, mineral oil, **206**:105

Sea birds, oil spill effects, **206**:104

Sea water pollution, petroleum products, **206**:95

Seafood residues, PFCs, **208**:180

Seasonal influences, HSP expression, **206**:15

Secondary plant metabolites, metal effects, **203**:141

Secondary poisoning, role in criteria setting, **209**:127

Secondary poisoning, wildlife dietary intake, **209**:90

Sediment concentrations, polycyclic aromatic hydrocarbons (table), **206**:104

Sediment cores, south San Francisco Bay contamination pattern, **206**:134

Sediment criteria, harmonization with aquatic values, **209**:132

Sediment detections, pharmaceuticals (table), **202**:96–97

Sediment effects, bioaccumulation, **204**:44

Sediment harmonization, in criteria setting, **209**:92

Sediment quality guidelines, Canada, **207**:46

Sediment residue, nonylphenol & its ethoxylate (table), **207**:44

Sediment residues, alkyphenol ethoxylates (table), **207**:42

Sediment residues, chlorinated paraffins (table), **207**:81

Sediment residues, flame retardants (table), **207**:75

Sediment residues, organic wastewater
    constituents (table), **207**:32–33
Sediment residues, pharmaceuticals, **207**:28
Sediment sampling, Great Lakes (illus.),
    207:13
Sediment sorption, pharmaceuticals, **202**:123
Sediment toxicity assays, gammarids, **205**:57
Sediment-oil interaction, marine environment,
    206:103
Sediments sampling locations, alkyphenol
    ethoxylates (illus.), **207**:39
Sediments, contaminant erosion at depth,
    206:131
Sediments, current-use pesticides (table),
    207:22
Sediments, methylmercury residue patterns
    (diag.), **206**:135
Seedling germination & growth, chromium
    inhibition, **210**:9
Sensitive species protection, setting criteria,
    209:124
Septicemia, *Pseudomonas* bacteremia, **201**:77
Serotonin reuptake inhibitors, description,
    202:67
Sewage epidemiology (forensics), in illicit
    drug source assessment, **210**:83
Sewage information mining, monitoring
    community-wide health, **210**:98
Sewage sludge contamination, chromium,
    210:5
Sewage sludge, pharmaceutical detections
    (table), **202**:104–108
Sewage treatment plant detections,
    pharmaceuticals (table), **202**:74–79
Sewage treatment plants, pharmaceuticals,
    202:72
Sex hormone effects, aging, **207**:125
Ship tanker accidents, oil spills (table), **206**:98
Ship tankers, petroleum transport, **206**:96
Shipping accident types, Baltic Sea (table),
    206:108
Shipping accidents & oil spills, Baltic Sea
    (table), **206**:107
Shipping accidents, Baltic Sea, **206**:106
Shipping oil spills, environmental effects,
    206:95 ff.
Shipping oil spills, implications, **206**:109
Silica, pulmonary toxin, **201**:55
Siliceous cell wall, diatom toxicity, **203**:90
Simulation modeling, estimating bioavailabil-
    ity, **203**:46
Single-species aquatic data, rating method
    (table), **209**:27

Single-species data, for criteria derivation
    (table), **209**:26
Skin cancer, PAHs, **203**:125
Skin effects, chromium, **210**:8
Skin infections, *Pseudomonas aeruginosa*,
    201:80
Skin-damaging effects, reactive oxygen
    species, **203**:122
Sludge land-use, pharmaceuticals, **202**:73
Soil adsorption values, OP pesticides (table),
    205:124
Soil adsorption, fenoxycarb (table), **202**:162
Soil adsorption, mechanism, **203**:15
Soil adsorption, PFOS (table), **202**:7
Soil aging effects, bound residues, **203**:25
Soil behavior, fenamiphos, **205**:135
Soil bioavailability, OP pesticides, **205**:129
Soil bioavailability, xenobiotics, **203**:1 ff.
Soil biodegradation pathway, 8:2-FTOH
    (diag.), **208**:167
Soil contact, Red Dog Mine contaminants,
    206:54
Soil degradation, coumaphos, **205**:150
Soil degradation, fenamiphos, **205**:134
Soil degradation, PCBs (diag.), **201**:146
Soil desorption, bound residues, **203**:20
Soil detections, pharmaceuticals (table),
    202:109
Soil fate, isofenphos, **205**:145
Soil fauna uptake, chemicals, **203**:12
Soil fauna, modeling bioavailability, **203**:57
Soil ingestion, human exposure, **203**:45
Soil metabolism, fenoxycarb (diag.), **202**:175
Soil microbe bioassays, bioavailability (table),
    203:31
Soil persistence, fenoxycarb (table), **202**:178
Soil persistence, fenoxycarb, **202**:171
Soil persistence, OP pesticides (table), **205**:130
Soil property affect, xenobiotic sorption,
    203:17
Soil residues, estimating bioavailability (table),
    203:61
Soil sorption effects, dissolved organic matter,
    203:20
Soil sorption effects, surfactants, **203**:20
Soil sorption effects, xenobiotic properties,
    203:17
Soil sorption processes, OP pesticides, **205**:123
Soil sorption, fenamiphos and metabolites
    (tables), **205**:136
Soil sorption, OP pesticides, **205**:122
Soil toxicity, mammals, **203**:41

Soil xenobiotics, desorption & dissolution, 203:19

Soils and xenobiotics, interactions, 203:15

Soils, chemical sorption isotherms, 203:15

Solid waste disposal, NTPs chemistry, 201:129

Solvent residues, surface waters (table), 207:31

Sorption isotherms, soils & chemicals, 203:15

Sorption processes, pharmaceuticals (table), 202:124–126

Sources of entry, PCBs (diag.), 201:140

Sources, perfluoroalkyl isomers, 208:115

Sources, TFA (table), 208:92

South Bay Salt Pond Restoration Project, San Francisco Bay (illus.), 206:119

South Bay Salt Pond Restoration, San Francisco water quality, 206:115 ff.

Species data requirements, aquatic criteria, 209:8

Species sensitivity distribution (SSD) methods, for calculating criteria, 209:38

Species variability, pesticide elimination, 204:40

Spilled petroleum, fish effects, 206:105

SSD (species sensitivity distribution) methods, for calculating criteria, 209:38

SSD failure, fit test, 209:111

SSD flow chart, in criteria setting (diag.), 209:111

SSD model comparison, discussion, 209:58

SSD percentile cutoff point, criteria derivation, 209:52

SSD procedure, checking goodness of fit, 209:109

SSD procedure, chronic criterion derivation, 209:114

SSD procedures, setting confidence levels, 209:53

SSD procedures, taxa aggregation, 209:54

SSD use, deriving acute criteria, 209:106

SSDs, UCDM context, 209:59

Statins, lipid regulators, 202:67

Stereoselectivity, RHOs, 206:81

Sterilization, using NTPs chemistry, 201:127

Steroids and hormones, description, 202:66

Steroids, organic wastewater contamination, 207:30

Stratospheric transport lifetime, HFCs, 208:26

Stress biomarkers, aquatic contamination, 206:4

Stress defense, fish HSP overexpression, 206:17

Stress hormones, fish, 206:6

Stress proteins, role in fish, 206:7

Stress response, heat shock proteins, 206:2

Stress, response in fish, 206:5

Stress, role for molecular chaperones, 206:6

Stressed fish, polluted environments, 206:1 ff.

Stress-induced production, HSP, 206:6

Stressor types, fish, 206:3

Stressors, impact on fish, 206:3

Structural nature, RHOs, 206:72

Structural residues, RHO pockets (table), 206:82

Structural types, OP pesticides (illus.), 205:120

Structures & properties, pharmaceuticals (table), 202:57–62

Structures, typical RHOs (illus.), 206:74

Styrene, pulmonary toxin, 201:49

Subsistence land use, lead exposure, 206:56

Substrate oxidation, RHOs, 206:78

Substrates, esterases (illus.), 204:57

Substrates, glutathione-S-transferase (illus.), 204:57

Substrates, oxidases (illus.), 204:57

Sulfonamides, antibiotics, 202:64

Sulfotransferases, pesticide metabolism, 204:73

Sulfur dioxide, pulmonary toxin, 201:48

Sunscreens, toxicity, 203:129

Surface runoff, mining wastes, 206:30

Surface water contamination, pharmaceuticals (table), 202:79–84

Surface water residue levels, pharmaceuticals (illus.), 207:28

Surface water residues, organic wastewater constituents (table), 207:31

Surface water residues, pharmaceuticals (table), 207:26–27

Surface water sampling locations, perfluorinated surfactants (illus.), 207:56

Surface water, *Pseudomonas aeruginosa* contamination, 201:89

Surface waters sampling locations, alkyphenol ethoxylates (illus.), 207:39

Surfactant contamination, North America, 207:37

Surfactant effects, chemical bioavailability (table), 203:22

Surfactant pollution, Great Lakes water & sediment, 207:38

Surfactant residues, fish, 207:40

Surfactant toxicity, gammarids, 205:7

Surfactants, soil sorption effects, 20

Survival role, HSP, 206:14

Sustainable remediation approach, phytoremediation, **206**:34

Svalbard glaucous gall breeding effects, oxychlordane residues (illus.), **205**:104

Svalbard glaucous gall effects, immunity and parasites, **205**:98

Svalbard glaucous gall metabolism, retinoids, **205**:93

Svalbard glaucous gall residues, accumulation affects, **205**:88

Svalbard glaucous gall residues, gonadal steroid hormones, **205**:95

Svalbard glaucous gall residues, hormone effects, **205**:94

Svalbard glaucous gall residues, reproduction effects, **205**:101

Svalbard glaucous gall residues, reproductive behavior effects, **205**:102

Svalbard glaucous gall residues, thermoregulatory effects, **205**:97

Svalbard Glaucous Gall, bioindicator species, **205**:77 ff.

Svalbard glaucous gall, chemical-induced effects, **205**:78

Svalbard glaucous gall, contaminant levels and patterns, **205**:79

Svalbard glaucous galls, bioaccumulation and age, **205**:87

Svalbard glaucous galls, brominated flame retardants, **205**:81

Svalbard glaucous galls, contaminant effects (table), **205**:89

Svalbard glaucous galls, contaminant genotoxicity, **205**:100

Svalbard glaucous galls, contaminants vs. next temperature (illus.), **205**:98

Svalbard glaucous galls, contamination, **205**:84

Svalbard glaucous galls, ecological biomarkers, **205**:89

Svalbard glaucous galls, gender and bioaccumulation, **205**:86

Svalbard glaucous galls, hydroxylated metabolites, **205**:82

Svalbard glaucous galls, mercury residues, **205**:86

Svalbard glaucous galls, metabolic enzyme effects, **205**:93

Svalbard glaucous galls, organochlorine residues over time (illus.), **205**:85

Svalbard glaucous galls, organometals, **205**:83

Svalbard glaucous galls, PBBs, **205**:81

Svalbard glaucous galls, PBDE residues, **205**:86

Svalbard glaucous galls, PBDEs, **205**:81

Svalbard glaucous galls, PFAS residues, **205**:83

Svalbard glaucous galls, PFOS residues, **205**:83

Svalbard glaucous galls, porphyrins, **205**:93

Svalbard glaucous galls, trace elements, **205**:83

Synergism, toxicity of mixtures, **209**:84

Synthesis pathway, fenoxycarb (diag.), **202**:164

T

Tailing wastes, iron-ore mining, **206**:30

Tailings from iron ore, generation process, **206**:31

Tap water, PFC contamination, **208**:199

Target organs, PFFAs, **202**:4

Taxa aggregation, criteria derivation, **209**:54

Taxa aggregation, SSD procedures, **209**:54

TCE (Trichloroethylene), dermal toxicity, **203**:128

TDIs (tolerable daily intakes), PFCs, **208**:202

Techniques, RHO degradation improvement, **206**:85

Temperature effects, criteria compliance, **209**:121

Temperature effects, bioconcentration, **204**:16

Teratogenic effects, isofenphos, **205**:148

Terrestrial data, role in UCDM, **209**:100

Terrestrial lab data, quality rating scheme, **209**:31

Terrestrial species toxicity, fenamiphos (table), **205**:141

TES (threatened & endangered species), criteria setting, **209**:88

TES, role in criteria setting, **209**:125

Test methods for gammarids, evaluating existing ones, **205**:62

Tetracyclines, antibiotics, **202**:65

TFA (trifluoroacetic acid) ubiquity, aquatic environment, **208**:92

TFA, direct & indirect sources (table), **208**:92

TFM-DS (10-trifluoromethoxy-decane-1-sulfonate), biodegradation pathways (dia.), **208**:172

TFMP-NS (9-[4-(trifluoromethyl)phenoxy]-nonane-1-sulfonate), biodegradation pathway (diag.), **208**:173

Theoretical approach, pesticide bioaccumulation, **204**:49

Therapeutic agents, pulmonary toxins, **201**:53

Therapeutic drugs, dermal effects, **203**:129

Thermal stability, PFOS, **202**:7

Thiol-containing molecules, HSF link, **206**:13

Threatened & endangered species (TES), criteria setting, **209**:88

Three-spot test, vacuuming dust contaminants, **201**:21

Threshold level effects, contaminants and avian species, **205**:104

Thyroid follicular cells, iodide uptake effects (table), **207**:123

Thyroxin ratio effect, PCBs (illus.), **205**:95

Tidal marsh habitats, south San Francisco Bay (illus.), **206**:118

Tillage effects, bioavailability, **203**:28

Time-response assay methods, gammarids, **205**:10

Tissue distribution, bioconcentrated pesticides, **204**:15

TNCB (2,4,6-Trinitro-1-chlorobenzene), toxicity, **203**:127

Tobacco, pulmonary toxin, **201**:56

Tolerable daily intakes (TDIs), PFCs, **208**:202

Toluene, pulmonary toxin, **201**:49

Tomato plants, iron-ore tailings treatment (illus.), **206**:36

Toxaphene toxicity data, comparative distribution fit (illus.), **209**:44

Toxaphene, data-set distribution test (diag.), **209**:41

Toxic action, electron transfer & oxidative mechanisms, **204**:135

Toxic effects, behavioral changes in MFB, **205**:27

Toxic effects, diatom nucleus & DNA, **203**:89

Toxic effects, dioxins & napthalenes, **203**:126

Toxic endpoints assessed, endocrine disruption, **209**:11

Toxicant effects, gammarid exposure modes (table), **205**:54

Toxicant exposure in infants, relative intensity, **201**:2

Toxicant skin penetration, aging effect, **207**:129

Toxicants in carpet, non-Hodgkins lymphoma, **201**:8

Toxicants, metals in dust, **201**:6

Toxicity assays, microbial activity, **203**:29

Toxicity attenuation, chromium bioremediation, **210**:1 ff.

Toxicity data fit to Burr III, pesticides (table), **209**:50

Toxicity data required, aquatic life criteria, **209**:7

Toxicity data summary, aquatic-criteria setting (table), **209**:21

Toxicity data, aquatic species required, **209**:8

Toxicity data, time parameters, **209**:14

Toxicity differences, PFA isomers, **208**:149

Toxicity effects on feeding, gammarid test methods (table), **205**:11

Toxicity effects, gammarid feeding activity, **205**:18

Toxicity endpoints assessed, aquatic life criteria, **209**:11

Toxicity endpoints, aquatic species, **209**:8

Toxicity endpoints, population-level effects, **209**:12

Toxicity estimation procedure, QSARs (table), **209**:126

Toxicity from oxidative mechanisms, examples, **204**:135

Toxicity implications, chemical mixtures, **209**:80

Toxicity in animals, fenamiphos (table), **205**:143

Toxicity in diatoms, siliceous cell wall, **203**:90

Toxicity in fish, fenamiphos and metabolites (table), **205**:140

Toxicity in mammals, fenamiphos, **205**:142

Toxicity in soil, mammals, **203**:41

Toxicity mechanism, electron transfer-reactive oxygen species-oxidative stress, **201**:42

Toxicity mitigation, antioxidants, **203**:131

Toxicity of chemicals, gammarids, **205**:7

Toxicity of colchicine, diatoms, **203**:89

Toxicity of formaldehyde, nervous system, **203**:105 ff.

Toxicity of mixtures, additivity models, **209**:81

Toxicity of mixtures, combining models, **209**:84

Toxicity of mixtures, concentration addition model, **209**:81

Toxicity of mixtures, response addition model, **209**:83

Toxicity of mixtures, synergism and antagonism, **209**:84

Toxicity reference values (TRV), PFBS, **202**:37

Toxicity testing, mining wastes, **206**:34

Toxicity to animals and plants, PFBS, **202**:22

Toxicity to gammarids, behavioral test methods (table), **205**:23

Toxicity value normalization procedure, water quality, **209**:122

Toxicity values, use with AFs, **209**:63

Toxicity, chromium, **210**:7

Toxicity, electrochemical interactions, **204**:137

Toxicity, formaldehyde, **203**:107

Toxicity, heavy metals, **206**:51

Toxicity, multiple-component effects, **203**:88

Toxicity, napthalenes, TNCB, nitric oxide
& carvoxime, **203**:127

Toxicity, OP pesticides, **205**:117 ff.

Toxicity, UCDM requirements, **209**:10

Toxicity, vs. bioavailability, **203**:4

Toxicokinetics, description, **203**:4

Toxicology, perfluorinated chemicals, **202**:1 ff.

Trace elements, Svalbard glaucous galls,
205:83

Transformation pathways, perfluorinated acid
precursors (diag.), **208**:13

Transformation, PCBs, **201**:142

Transport in environment, OP pesticides,
205:121

Transport mechanisms, chemicals, **203**:5

Transport processes, OP pesticides, **205**:122

Transport role, chromium complexes, **210**:6

Transport, illicit drugs, **210**:94

Tree species, iron-ore tailings uptake (illus.),
206:36

Trichloroethylene, pulmonary toxin, **201**:49

Triclosan detections, Great Lakes waters,
207:35

Trifluoroacetic acid (TFA), atmospheric
reactions, **208**:82

Trophic level effects, PFCs, **208**:193

Trophic magnification, perfluorinated
compounds, **207**:53

TRV , PFOS (table), **202**:34

TRV, PFBS (table), **202**:37

TRV, PFBS, **202**:37

TRV, PFOS (table), **202**: 34

U

UCDM (University of California-Davis
Methodology), water quality, **209**:3

UCDM context, SSDs, **209**:59

UCDM development, data sources (table),
209:17

UCDM development, web-address sources
(table), **209**:19

UCDM goals, water quality criteria, **209**:5

UCDM requirements, acute & chronic toxicity,
209:10

UCDM, acute criterion derivation, **209**:105

UCDM, aquatic toxicity data, **209**:98

UCDM, assumptions & limitations, **209**:93

UCDM, bioaccumulation & food residues,
209:89

UCDM, chlorpyrifos aquatic criteria values
(table), **209**:137

UCDM, criteria derivation flow chart (diag.),
209:97

UCDM, data collections details, **209**:98

UCDM, data flow chart (diag.), **209**:97

UCDM, data requirements, **209**:6

UCDM, evaluating ecotoxicity data, **209**:101

UCDM, evaluating physical-chemical data,
209:101

UCDM, goals & definitions, **209**:95

UCDM, population-level endpoints, **209**:13

UCDM, uncertainties, **209**:93

Uncertainties, UCDM, **209**:93

University of California, water-quality
methodology, **209**:1 ff.

University of California-Davis Methodology
(UCDM), water quality, **209**:3

Uranium, aquatic species behavior effects,
210:46

Uranium, brain effects, **210**:39

Uranium, human behavioral response, **210**:42

Uranium, rodent behavioral studies, **210**:44

Urea pesticides, chemical structures (illus.),
204:102

Urinary system changes, aging effects, **207**:98

Urinary tract infections, *Pseudomonas
aeruginosa*, **201**:78

US stream residues, current-use pesticides
(table), **207**:20–21

US surface waters, pesticide detections, **207**:19

USEPA, water quality methods, **209**:2

Uses, chromium, **210**:5

UV light, *Pseudomonas aeruginosa*
disinfection (table), **201**:96

V

Vacuum cleaners and contaminants, cleaning,
201:20

Vegetation, isofenphos breakdown, **205**:146

Ventricular performance, aging effects,
207:105

Vertebrate metabolism, PCBs and PBDEs,
205:82

Vertebrates, metabolic enzyme effects, **205**:93

Veterinary drugs, illegal uses, **210**:92

Viruses, in house dust, **201**:18

Vision effects, aging (table), **207**:113

Vitellogenin-like proteins, gammarid
biomarkers, **205**:45

VOC pollutant removal, NTPs chemistry,
201:123

Volatile anesthetics, atmospheric lifetime,
208:13

Volatile precursors, yielding perfluorinated
acids (table), **208**:93

Volatility, OP pesticides, **205**:124
Volatilization of PCBs, environmental entry, **201**:140
Volatilization, PCBs, **201**:146

W

Waste description, iron-ore tailings, **206**:31, **206**:2
Waste utilization from mining, phytoremediation, **206**:34
Wastes from iron-ore mining, characteristics, **206**:30
Wastes from mining, phytoremediation, **206**:29 ff.
Wastewater irrigation, pharmaceutical release, **202**:73
Wastewater treatment plants, pharmaceutical fate, **202**:111
Wastewater treatment, NTPs chemistry, **201**:125
Wastewater treatment, pharmaceuticals (table), **202**:112–117
Water & sediment detections, pharmaceuticals, **207**:25
Water chemistry effects, bioconcentration, **204**:16
Water contamination, PFCs, **208**:180, 199
Water pesticide contaminants, Great Lakes, **207**:15
Water quality & San Francisco Bay, future changes, **206**:138
Water quality criteria use, physical-chemical data, **209**:22
Water quality criteria values, PFCs in aquatic species (diag.), **202**:25
Water quality criteria, methodology goals, **209**:5
Water quality criteria, pesticides, **209**:1 ff.
Water quality criteria, PFCs, **202**:22
Water quality criteria, PFOS & wildlife, **202**:33
Water quality criteria, UCDM, **209**:3
Water quality effects, criteria compliance, **209**:117
Water quality effects, on toxicity, **209**:76
Water quality effects, role in UCDM, **209**:100
Water quality methods, USEPA, **209**:2
Water quality standards, *Pseudomonas aeruginosa*, **201**:101
Water quality testing, *Gammarus* spp., **205**:1 ff.
Water quality, south San Francisco Bay, **206**:115 ff.
Water quality, south San Francisco Bay, **206**:120

Water quality, toxicity value normalization procedure, **209**:122
Water residues, alkyphenol ethoxylates (table), **207**:42
Water residues, chlorinated paraffins (table), **207**:81
Water residues, fenamiphos (table), **205**:135
Water residues, flame retardants (table), **207**:75
Water residues, from PFCs (table), **208**:200
Water residues, nonylphenol & its ethoxylate (table), **207**:44
Water residues, pesticide detections, **207**:19
Water residues, synthetic musks (table), **207**:50
Water sampling, Great Lakes (illus.), **207**:13
Water solubility, hydrocarbons from petroleum, **206**:99
Water-induced degradation, fenamiphos, **205**:134
Watershed map, San Francisco Bay (illus.), **206**:117
Weathering, fate of oil spills, **206**:100
Web-address sources, UCDM development (table), **209**:19
Wetland restoration, San Francisco Bay, **206**:117–118
Wildlife contamination, methyl mercury, **206**:123
Wildlife contamination, perfluorinated compounds, **207**:55
Wildlife dietary intake, secondary poisoning, **209**:90
Wildlife effects, fenamiphos, **205**:141
Wildlife effects, lead exposure, **206**:56
Wildlife exposure to lead, Red Dog Mine, **206**:56
Wildlife exposure, Red Dog Mine, **206**:51
Wildlife residues, PFOS isomers (table), **208**:144
Wildlife residues, San Francisco Bay, **206**:125
Wildlife water quality criteria, PFOS, **202**:33

X

Xenobiotic bioavailability estimation, chemical extraction (table), **203**:43
Xenobiotic clearance, aging effects, **207**:101
Xenobiotic effects, diatom cell wall, **203**:90
Xenobiotic properties, soil sorption effects, **203**:17
Xenobiotic soil occlusion, bound residues, **203**:26
Xenobiotic sorption, soil property affect, **203**:17
Xenobiotic uptake by roots, influencing factors, **203**:12

Xenobiotic uptake, higher plants, **203**:34

Xenobiotics and soils, interactions, **203**:15

Xenobiotics in soil, desorption & dissolution,
203:19

Xenobiotics, agents in immunotoxicity (table),
207:121–122

Xenobiotics, aging effects, **207**:109

Xenobiotics, soil bioavailability, **203**:1 ff.

Xenobiotics, toxicity vs. bioavailability, **203**:4

Y

Yeast, chromium bioremediation, **210**:15

Z

Zinc & lead concentrates, Red Dog Mine,
206:50

Zinc exposure, humans, **206**:51

α Subunit structure, catalytic pocket (illus.),
206:78

α Subunit structure, RHOs, **206**:75–78

α-Dicarbonyl species, advanced glycation end
products, **204**:142

α-Dicarbonyl species, reactive oxygen
& carbonyls, **204**:142

α-Dicarbonyls, bioactive examples, **204**:143

α-Dicarbonyls, diacetyl relationship, **204**:137